# How to Cut Costs and Improve Service of Your

TELEPHONE
TELEX
TWX
and other TELECOMMUNICATIONS

Frank Kern Griesinger

President, Frank K. Griesinger and Associates, Inc.

# How to Cut Costs

# and Improve Service of Your

## TELEPHONE
## TELEX
## TWX
## and other TELECOMMUNICATIONS

McGRAW-HILL BOOK COMPANY

New York St. Louis San Francisco Düsseldorf Johannesburg
Kuala Lumpur London Mexico Montreal New Delhi
Panama Paris São Paulo Singapore
Sydney Tokyo Toronto

Library of Congress Cataloging in Publication Data

Griesinger, Frank K
How to cut costs and improve service of your telephone, telex, TWX, and other telecommunications.

1. Telephone in business. I. Title.
HE8735.G74 651.7 73-21616
ISBN 0-07-024760-9



234567890 KPKP 7654

*The editors for this book were W. Hodson Mogan and Beatrice Carson, the designer was Naomi Auerbach, and its production was supervised by Teresa F. Leaden. It was set in Alphatype Astro by University Graphics, Inc.*

*Printed and bound by The Kingsport Press.*

*I dedicate this book to two men who have had a major role in my study of the telecommunications field:*

*ROBERT B. SEIDEL, who suggested in 1958 that clients of my consulting corporation might wish some advice on negotiating with the communications utilities and who worked with me in developing our first written instructions*

*and*

*COLONEL ROY STONE, the first professional consultant in the field, who helped me develop this division of our management consulting practice when we began this work in 1960*

# CONTENTS

# PREFACE

Have you been wondering how to cut the cost of long distance calling? Are there steps you can take to reduce the amount you pay on your monthly bill for message units? Should you be investigating the possibility of installing telephone equipment which you can purchase or rent from a private vendor and interconnect to your utility's network facilities? Can your costs be reduced substantially if you use modern written message services such as Telex, TWX, and Mailgram?

Consumers of telephone and telegraph services have needed a guide which would help them understand the services and equipment furnished by the telephone and telegraph utilities and by the independent manufacturers of telephone equipment. This book has been written to fulfill that need.

Here you will find suggestions for reducing cost, improving service to your customers and employees, and enhancing your organization's public relations image through the proper use of your telecommunications facilities. Have you ever read the public tariff which governs the rates and services furnished by your utilities? Examples of how to interpret this information are given in the chapter on tariff reading. If you supervise telephone and written wire communications services in your headquarters location, you'll find suggestions on making the most efficient use of the facilities you have. Equally important, you'll read about the factors you should analyze if you are considering private systems as alternatives to the systems proposed by the utilities. For those of you whose management responsibility extends to the branches and divisions of your company, there are helpful ideas on designing telephone systems for servicing localities with many people or with just a few employees. If you're responsible for a small system, I think you will find that many of the suggestions contained in this book apply to your problems just as much as they do to those of larger organizations. Finally, even if you are an experienced telecommunications professional, you may profit by some of the ideas and forms in the chapters on accounting and organization.

The knowledge you gain from this book will make it easier for you to talk with representatives of the communications utilities and with vendors of interconnected facilities and equipment. You will be speaking the language of the industry. Your expertise will help you secure the services you need at reasonable cost.

This book is not intended to be a critique of the telecommunications industry. While some utility marketing representatives and interconnect vendors may have sold the "top of the line" without identifying lower-cost alternatives, the majority of the industry's marketing people are dedicated to serving the customer well. As a management consultant and as a former officer of a major manufacturing organization, I have worked with these individuals throughout my business career and deeply respect their integrity, intelligence, and attention to their users' needs and budgets.

Telecommunications technology changes rapidly. This book contains the latest available information, but you should check with your utility or private vendor for current information on costs and services. To ensure that my text is accurate, I asked members of the industry to read selected chapters. I thank them for their valuable advice. Several individuals at Western Union's headquarters offered excellent suggestions which improved Chapter 14. The Public Affairs department of RCA Global Communications, Inc., made helpful comments on Chapter 15. I am grateful to Bruce Walker of the Howard Johnson organization and to my engineering associate, Andrew Bass, for splendid assistance on the vital subject of interconnection, described in Chapter 12.

Many of the thoughts incorporated in this book have come from the men and women who attend my seminars on telecommunications. Similarly, my management consulting practice has brought me into contact with talented people who manage telecommunications in organizations located throughout the United States and Canada. Any management consultant learns a great deal from those with whom he works, and I must credit many ideas in this book to members of my client group.

Finally, I want to express my appreciation to Bettie Wiechelman and Louise Figer of my own corporation for their work in preparing the manuscript. My editor at McGraw-Hill, W. Hodson Mogan, has given much helpful advice. And I am especially grateful to Beatrice Carson of McGraw-Hill for her fine editorial assistance.

As you may be able to tell when you read this book, I am intrigued and excited by the constantly developing technology which permits us to send the human voice, the written word, and bits of data quickly around the world. I hope my enthusiasm for the subject of telecommunications will inspire you to further enlarge your knowledge of this fascinating industry.

*Frank Kern Griesinger*

CHAPTER ONE

# MANAGING LONG DISTANCE TOLL EXPENSE

Cost-conscious executives constantly urge managers to cut the telephone bill. But seldom does management guide the staff on how cost-cutting can be done. This chapter gives suggestions which can bring about cost reductions on long distance expense ranging from 42 to 76 percent on a three-minute call. To accomplish this, you and your associates should experiment with the guidelines you are about to read and then incorporate them into your organization's general policy. Once you establish standards, you can analyze your billings for long distance expense. By identifying the called numbers, you can determine which departments need additional education in directly dialing station-to-station calls.

Almost every business person *wants* to help his management reduce cost and improve profit. The methods which follow are "good sense" ideas. You'll be able to "sell" them easily! Also, your people will immediately perceive that they can save money on their residence long distance bills by following these suggestions.

Try to adapt this chapter to any special conditions existing in your organization. Write a memorandum of instructions to your people, and watch your cost per call decline! In the paragraphs which follow, you'll notice that I am talking to *you*—and when you write your memo, try to talk enthusiastically and personally to your coworkers. Remember: these are proven suggestions which save money and executive time. They will work for you, too!

## INTERSTATE LONG DISTANCE CALLS[1]

The lowest-cost call is the call you directly dial yourself from your office or residence telephone, or one which your switchboard operator dials, without asking for assistance from the operator of the telephone company (telco). This is known as Direct Distance Dialing (DDD). Under normal circumstances, the called telephone will be ringing in less than 30 seconds. The *local time* when you place your call has a lot to do with the price you'll pay. So does the class of call you place: "station," "person," or "operator-handled." Let's look at some examples. The dollar rates used in this chapter were those in effect in late 1973. The federal excise tax must be added to the rates quoted, plus any applicable state or local taxes. In 1974 the federal excise tax is 8 percent. It is scheduled to be reduced 1 percent per year.

### DAYTIME DDD CALLS: MONDAY–FRIDAY, 8 A.M.–5 P.M.

This rate period is the normal business day. If you directly dial your call, the three-minute rate cannot exceed $1.45 to any point within the United States. In contrast, if you dial telco's operator and place a person-to-person call, the rate jumps to $3.55. Why should you use a "person" call when you can make two DDD calls for less than the price of one "person" call?

*Here's the action you can take* to impress your boss with a 59 percent cost reduction on a three-minute call. Find out *who* is placing the "person" calls in your organization. Your monthly toll bill lists the calls. Usually you'll find a code which tells you if the call was station or person. If the code doesn't appear, call your telco business office and ask them to send you the stub end of your bill from their files. The stub will give you the length of call, type of call, and time the call was placed. You are *entitled* to receive this information whenever it is needed.

Next, ask the business office to try to identify those who placed person-to-person calls from your office. This may not be possible, because modern telephone exchanges utilize an operator position which features "zero-plus" dialing. In such exchanges, the person placing an operator-handled call is instructed to first secure an outside trunk. Then "zero" is dialed, followed by the area code and telephone number. This information appears on a screen at the

[1] The Bell System has recently renamed the interstate long distance service offering which is listed in AT&T Tariff 263. The new name is Message Telecommunications Service (MTS).

utility operator's position. She takes your name and the name of the person you wish to reach, and she stays with you until you reach your party, talk to an acceptable alternate, or leave call-back instructions. In such exchanges, the name of the caller may be given to the called person, but it may not be recorded in telco's magnetic tape or punched cards. Similarly, the name of the called person may or may not be available on telco's records. However, your telco can *always* identify the individual or organization associated with the called telephone *number*. Ask your telco to give you all the information possible. If you can identify your associates who are making person-to-person calls, tell them about the economies of direct dialing on a station basis. Ask for their help. Then, keep checking your monthly long distance statements to make sure that your educational work is effective.

### EVENING DDD CALLS: MONDAY–FRIDAY, 5–11 P.M., SUNDAY, 5–11 P.M.

You've done your educational work on the regular business day. Your people are using DDD. Would a further cost reduction of 42 percent sound good to your management? Remember that people in time zones west of you are still at their desks when you're about to go home at 5 P.M. Could you wait a minute and place your call just after 5? Now that $1.45 call will cost only 85 cents. At 5 P.M. in New York City it's only 2 P.M. in Los Angeles, 3 P.M. in Denver, and 4 P.M. in Chicago.

*Here's the action you can take:* Remind your department heads that the three-minute rate declines by 42 percent at 5 P.M. Ask them to defer calls to Western points until that time. The network is less crowded then, so calls go through faster. Also, if your people are working on order solicitation or credit and collection, using lots of long distance service, ask them to consider breaking down their calling lists into geographic segments. Then, perhaps your solicitors could come to work two or three hours later and stay after 5 P.M. to concentrate Western calls in the bargain toll-rate period. After-5 calling is particularly useful if you telephone people at their homes.

### WEEKEND DDD CALLS: SATURDAY, 8 A.M.–11 P.M.; SUNDAY, 8 A.M.–5 P.M.

Do any of your people work on Saturdays? How about your customers? Do you have kids away at school whom you'd like to call? The DDD network is lightly used on weekends, so you are given a real communications bargain. That $1.45 call now costs only 70 cents—a whopping 52 percent cost reduction from the Monday–Friday rate!

*Here's the action you can take:* See if any of your customers or vendors customarily work on Saturdays. If so, consider scheduling your order solicitation crew for a Saturday shift. If you do business with individuals at their homes, try making your phone calls on Saturday. If you want to chat with kids, relatives, and friends, call them on Saturday or Sunday. After 5 P.M. on Sunday the evening rate schedule takes effect.

### NIGHT DDD CALLS: DAILY, 11 P.M.–8 A.M.

Do you have offices in the West? Do you have some eager managers or salesmen who don't mind doing their telephoning early? Any DDD call which is connected prior to 8 A.M. daily takes the lowest rate of all the rate periods, and a *one-minute* minimum applies, rather than three minutes. The coast-to-coast rate is only 35 cents; a call traveling 100 miles costs only 15 cents for that first minute. Additional minutes are even less. And as long as your call is answered before 8 A.M., you can talk *all day* at the bargain rate! If you can complete your business in a single minute, you've made a 76 percent cost reduction over the daytime rate!

*Here's the action you can take:* Encourage your Western people to directly dial calls before 8 A.M. from their office or residence, perhaps sending you their residence toll statement for reimbursement. Consider connecting one of your office trunk lines to a recording machine when your switchboard closes. Release that special number to your people in distant cities so that they can call you after 5 P.M., preferably after 11 P.M., and surely before 8 A.M. to leave requests for information, record orders, or perhaps brief you on problems which can be handled by an appointment toll or Wide Area Telecommunications Service (WATS; see Chapter 3) call the following day. The possibilities are great—just use your ingenuity!

### OPERATOR-HANDLED CALLS

When an operator must assist in some aspect of a long distance call, a charge is made for her labor. The telephone utilities are gradually mechanizing the handling of operator-assisted calls. In modern technology, the operator is located at a console which is part of the Traffic Service Position System (TSPS). You dial "0" for the utility's operator. You then dial the area code and the desired telephone number. These digits appear on a screen in front of the operator. She says something like: "May I help you?" After you have given her your special instructions, she presses a key which releases your call to the network, and you will hear the ringing of the distant telephone.

Operator-handled calls are always billed at higher rates than calls which you dial directly. In late 1973 a flat charge was made for operator handling. During business hours it ranged from 3 cents to 40 cents per three-minute call, the charge varying with distance. On calls traveling less than 244 miles, an extra charge of 1 to 3 cents is made for each minute of overtime. After 5 P.M. on Monday through Friday, the cost of operator handling is even greater, as basic rates are reduced. During the 5 P.M.–11 P.M. rate period, the cost for operator handling on a three-minute call varied from 7 cents to 55 cents in 1973. The federal excise tax applies to the cost of operator handling, as do state and local taxes.

It is important that you know the types of calls which are billed at higher rates and what action you can take to avoid the rate penalty for operator handling.

## Credit Card Calls

Telcos make no charge for issuing credit cards. However, when you dial and give your credit card number to the operator, the special surcharge is applied.

*Take action:*

1. Don't use your credit card at home. Directly dial your call, then claim reimbursement on your regular expense account at month's end. This assures the lowest rate for your employer or for you.
2. If *you* are an employer and have issued credit cards to your employees, consider recalling them. Ask your people to apply for a *personal* credit card. Both cards work the same way, but if your employee terminates his employment, you won't have to pick up his card. Also, he will make fewer personal calls at company expense. Business calls are listed on his expense report.
3. If you are calling from a coin station, you'll save time by using your credit card. Toll operators increase or decrease odd-penny toll rates to the nearest nickel. Your credit card billing will show the *actual* rate.
4. If you are calling from a hotel or motel phone, *do* use your credit card. Depending on the state, the *intrastate* toll call made

on a credit card may be less expensive than a call charged to your room. This is because hotels are permitted to add a service charge to intrastate toll calls. The charge may be less if your credit card is used, depending on the tariff of the local utility.

### Calls Made from Coin Telephones

These calls are always considered "operator-handled" even if you do all the dialing yourself and the operator only says at the end "May I help you?"

*Take action:* Use your credit card, if you have one. It will ensure accurate billing for your call.

### Third-Number Calls

You want to make a call to your office from a customer's plant. You dial the telco operator and instruct her to charge your office telephone number for the call. Such calls are "third-number" calls and are subject to the surcharge.

*Take action:* If you now collect cash from employees or make payroll deductions to cover the cost of their personal calls, save yourself all that paperwork by instructing the employee or your switchboard operator to have the call charged to the employee's residence telephone.

### Collect Calls

Many companies encourage customers to call collect. Traveling employees may make collect calls. Collect calls are always subject to the surcharge, and unless the calling party is carefully educated on procedure, you probably will be paying for a very expensive person-to-person collect call.

*Take action:*

1. Acquaint your customers and employees with the fact that if they ask for John Jones, or for the order department, or even give the operator your extension number, they are making a *"person"* collect call at more than double the "station" rate.

2. As explained in an earlier paragraph, your telco may be able to give you information on the identity of individuals who are making collect person calls to you. If this information is available, it will help you in your educational work on the use of direct dialing and of station collect calling.

3. Have your sales department write diplomatically to your customers, reassuring them that you are delighted to have them call collect but asking them to place their call for "anyone" at your

main number. Then assure them that your switchboard operator will immediately accept the collect call and will find a qualified person to take immediate action on the order or request. Be sure to give your operator a list of "approved" collect callers.

4. Do not permit your own employees to place person-to-person collect calls to your office. If you wish to accept collect calls from your employees, instruct them to tell the utility's operator that they will talk to "anyone" at your telephone number. Your switchboard operator will then accept the call and connect it to the wanted person or department. Be sure to tell your employees that if they are calling from their home they should dial direct. They then submit their residence telephone toll statement as evidence that they made a DDD business call on their residence telephone.

**Refusing Collect Calls.** Some companies instruct their own employees, dealers, or agents to call collect. The receiving switchboard operator hears a statement from telco's operator like this: "I have a collect call for anyone (or for Mr. John Jones) from Mr. Brown. Will you accept charges?" The switchboard operator is told to refuse the collect call. As she now knows the identity of the calling person, which may be a familiar name, she can pass that information along to an appropriate person. He then calls back in the least expensive way: via DDD, WATS, foreign exchange (see Chapter 4), or tie line (Chapter 5).

This procedure is a tariff violation. Consequently, the telcos have developed a special computer program to catch the violators. The computer lists all refused collect calls by originating area code and number. It then matches outgoing calls from the called number to the originating number. The offending organization is then told to stop this practice or risk losing its phone service.

However, the telephone utilities recognize that they save labor expense for operators when a customer uses the DDD network. Consequently, operators have been instructed to release the telephone number of the individual calling collect if the called person is not available. Thus, in legitimate situations when such a circumstance occurs, you can secure the distant number. When your absent person returns he can place the call, using the least expensive method available.

## Special Toll Billing

Some telcos offer "special billing numbers" for use at an organization's principal office. The service is sometimes known as "code billing" or " QZ billing." A charge is made for each group of numbers, usually with a minimum rental charge for the first 50 numbers. The per-number rental cost averages 15 to 25 cents monthly. When this service is adopted, your ninth-level (dial "9") outgoing trunks are wired so that an intercepting operator comes on the line at the *end* of your dialing. She asks for your number. Instead of giving her your main telephone number, you give her the special billing number. At the end of the month you receive a separate toll statement for each number, thus permitting you to make an accounting distribution on a departmental or an individual basis. Except in a very few localities which do not have modern central office switching equipment, these calls are now billed at the higher operator-handled rate. Most telcos are deemphasizing this service, and some have discontinued it entirely because operating cost is greater than revenue.

*Take action:* If you are using this system, management should realize that your toll costs are now 15 to 20 percent higher than they would be if your people used the DDD procedure. In addition, you are paying a monthly fee for the billing codes. You have several alternative choices:

**Issue Credit Cards.** Your telco may be willing to issue individual credit cards and to permit use of those cards at your main location. This procedure will save the rental charge for the special billing numbers, but calls will still be billed at the operator-handled rates. Also, call completion will be slower, as you will have to dial telco's operator and wait for her to answer and to dial for you.

**Prepare Individual Toll Tickets.** This alternative is difficult to enforce, but it is better than handing an unrestricted calling privilege to your people. You prepare pads of call tickets. As your person waits for telco's automation to speed his call over the DDD network, he puts the date, area code, telephone number, and his name and cost center on the ticket. You may also want to have him record the name of the called company or individual to better protect you against personal calls. These tickets are sent to your chief operator or to your accounting department, and they are checked against the toll bill after being sorted by date.

**Identify Personal Calls.** There will always be many calls for which no call ticket has been received. These calls can be reviewed. When there are many calls to the same number, or when calls are very expensive, the auditing person records the date, area code, and telephone number on special forms which are furnished by most telcos. The business office of telco sends back the form with all available information. The name of the call originator and receiver may be shown on any "person" call. The name of the distant originator may be available on any "station" collect call. The name of the telephone subscriber at the distant end is available on a directly dialed call or an operator-handled station call. With this information, a knowledgeable person can make assumptions on the identity of the calling person or department. Where there seems to be flagrant abuse, telco's business office may be willing to call the distant phone and see if they can determine the identity of your calling person.

In Chapter 3, I will describe a special computer study that lists toll calls by area code, exchange, and telephone number. All calls to a specific telephone number are listed in one section of the study. The distant number can be identified, and if it is not a known customer or vendor, telco's business office may be able to investigate. A telco representative calls that number and tries to determine which of your employees has made those calls.

If such a computer study cannot be made available to you, perhaps your own data processing department can assist you. Almost every telco is willing to sell or rent its punched cards or magnetic tapes from which its toll billing is prepared. You secure this input material and instruct your data processing staff on the "sort" you want. They can then give you a computer printout which will analyze your toll calls in any format which will be helpful to you.

**Install Semirestricted Telephones.** For a one-time installation charge, telco can restrict a particular station in your dial system so that the user receives a busy signal when he dials "9." This means that all local calls and all toll calls must be handled through the operator or else made at an unrestricted station. This arrangement is useful chiefly for preventing clerical personnel from making personal local and toll calls on company facilities.

**Install Toll-Diverting Apparatus.** A special switch is installed by telco on each outgoing ninth-level trunk. If the caller dials "9" and then an exchange or area code which is outside the local calling area, the call is diverted to the switchboard operator, or the caller may receive a "busy" tone. The dialing of a long distance access code, such as "1," will also divert the call. If the caller dials "9" and then "0" for the utility's operator, the call is likewise diverted. Some sophisticated diverting equipment may permit the caller to dial certain exchanges in the local message unit area, but not other exchanges where multiple message unit charges apply.

The monthly cost of this equipment varies widely among the utilities, but you may find that the cost ranges from \$3 to \$12 per trunk equipped. The equipment must be installed on every ninth-level outgoing trunk, not just on selected trunks.

If you install toll diversion, you must dial your operator to make a long distance call. She may prepare a toll ticket for accounting purposes. She may dial the call for you or extend the dial tone so that you can do the dialing yourself. If toll calling is heavy, one or more switchboard operators may have to be added to handle the additional workload. Thus, you must weigh the cost of equipment rental and switchboard labor against the monthly cost of the unidentified calls which appear on your long distance bill. Toll diversion is not widely used, but in a large company which cannot yet justify expensive automatic toll-ticketing equipment, toll diversion may force personnel to use lower-cost WATS, foreign exchange, or tie-line circuits, as the operator can route the call over the lowest-cost circuit available.

**Install Automatic Toll-Ticketing Equipment.** Many vendors now offer interconnected equipment which will identify the extension (station) number from which any toll, WATS, foreign exchange, or tie-line call originates. A computer-type printout is created which also identifies which outgoing line was used, the area code and number dialed, the time of day the call was placed, and the duration of the call in minutes and tenths of minutes. While the purchase or lease cost of these units is substantial, a major organization may find that the monthly rental cost is appreciably less than the cost of personal calls which may be made on an unrestricted telephone system.

### Zenith and Enterprise Service

These terms mean the same thing. Telco assigns a special four-digit number to you, preceded by the word "Zenith" or "Enterprise." You release this number to the customers or employees who are authorized to call you toll-free. If you wish, the number can be listed in the telephone directory serving a particular geographic area from which these calls will originate. A separate monthly rental charge is made for the listed or unlisted number in each such area. That charge is usually in the $3 to $6 range. Your customer dials his telco operator and asks for the Zenith or Enterprise number. The operator directly dials the call to your switchboard, and the customer speaks to the switchboard operator immediately without any introduction from telco's operator. The call is billed at the operator-handled rate, as it is considered a "collect" call. Also, if the caller asks the operator for a specific person, department, or extension number, your Zenith or Enterprise call may be billed at the person-to-person collect rate.

*Take action:* Find out if your organization is using this service, either at headquarters or at branch points. If so, make sure management knows that, in addition to paying the operator-handled surcharge, you must pay $36 to $72 or more yearly for the rental charge of the Zenith or Enterprise number. If this number is listed in several directories, a separate rental charge applies for each directory. Also, telco usually does not itemize each such call received but bills them in a lump sum, merely showing the *number* of calls received on a particular Zenith or Enterprise number. There are two alternatives. Your customers can be told to call "collect" for "anyone" at your main switchboard number; this saves the monthly rental charge for the special number. Or you can order inward WATS service and release the "800" number to those for whom you wish to pay toll charges. This service is described in Chapter 3.

### "Time and Charges" Calls

Hotels and other organizations that require quick notification of the cost of a long distance call may ask the utility's operator for "time and charges." After the call is completed, the utility's operator calls the switchboard and states the length and cost of the call. Or, in larger establishments, the time and charge information may be placed on an incoming teleprinter. Such calls are considered operator-handled calls.

*Take action:* You can prepare a call ticket for the call. At the end of the month you can secure the charge from the toll statement mailed with your telephone bill.

## INTERSTATE CALL OVERTIME CHARGES

The interstate rate schedule effective in late 1973 gives a bonus to the long talker. But advocates of consumerism may well wonder about AT&T's method of calculating the rates applicable to the various mileage zones. The average discount for each overtime minute is 11.3 percent compared to the rate per minute for the initial three-minute period. However, the discount ranges from 20 percent for calls traveling 17 to 22 miles to 5 percent for calls traveling 1,911 to 3,000 miles. In general, the larger discounts apply to the shorter distances.

Formerly, the surcharge made for operator-handled calls applied only to the first three minutes. Charges for overtime minutes were the same as the charge made for overtime on DDD station calls. For reasons unknown, AT&T now charges 1 to 3 cents more per overtime minute for operator-handled calls which travel less than 100 miles. The call traveling 149 to 196 miles costs an extra 2 cents per overtime minute. The call traveling 197 to 244 miles costs 1 cent extra per overtime minute. Person-to-person overtime rates are handled exactly the same as operator-handled rates.

During the rate periods other than Monday–Friday from 8 A.M. to 5 P.M., even greater surcharges apply to overtime on both operator-handled station calls and person-to-person calls. Two-cent to six-cent surcharges are made per minute of overtime in certain mileage zones. The basic rate for the first three minutes of person-to-person calling remains the same at all rate periods. Person-to-person overtime charges follow the pattern and rate of the operator-handled call.

## LONG-DISTANCE CALL-BACKS

To speed handling of call-backs on incomplete person-to-person toll calls, all telcos now leave a message with the answering person giving the name of the calling party, the area code and number, and an instruction to call "Operator 6" or "Operator 7." "Operator 6" identifies a regular person-to-person call. "Operator 7" identifies a special billing call, as described earlier in this chapter. Before she connects your person with the person who called earlier, she will ask him for his special billing instructions.

*Take action:* If you dial telco's operator, then ask for Operator 6 or 7, and let telco complete the call, the person who called you will be billed at the high person-to-person rate. Thus, if you recognize the calling person's name as one of your own organization's people, ignore the call ticket. Instead, place a directly dialed call to that number. The person is waiting to hear from you. You should be able to reach him via DDD on your first attempt. There is never any charge for the lapsed Operator 6 or 7 call, as most telcos destroy their charge ticket as soon as they "leave word."

On the other hand, if your calling party is recognized as a stranger, vendor, or customer, *do* call telco's operator and have her complete the call. Such calls are paid for by your calling party, and thus they will not appear against your budget!

## PROBLEMS WITH THE DDD NETWORK

If you reach a wrong number, ask the answering person the name of his city and his telephone number. Dial telco's operator and ask her to credit you for that call. She will do so without question. You can dial again.

If you make several attempts to directly dial a call and find that the circuits are busy or that you hear the recording "I am sorry, but your call did not go through: please try again," dial telco's operator for assistance. Explain your several attempts. She will dial the call for you and will record details on a special form which will give you the DDD rate, not the operator-assisted rate. Caution: Be *sure* to emphasize you've tried several times, and, as a double check, ask the operator to make sure you are billed at the low DDD rate.

## INTRASTATE LONG DISTANCE CALLS

In the first part of this chapter we discussed interstate toll calls. Rates for such calls are regulated by the Federal Communications Commission (FCC). In contrast, an intrastate rate is set by the telcos operating in that state. The rate is then regulated by the state public utility commission or, in Texas, by city government authorities. The rate-approving bodies have formed the National Association of Regulatory Utility Commissioners. I suggest that you order the association's two publications.[2]

One booklet, *Long Distance Message Telephone Rates,* is published yearly. It contains tables showing airline mileage groups in every state and the rate applicable in each group for every type of intrastate call. Also, the booklet gives the interstate rate schedule in the same airline-mile format. Rates are also shown between the United States, Canada, and Mexico. Furthermore, you should obtain a copy of the *Official Airline Guide,* in which you'll find a table of airline miles between principal cities. With the booklet and the airline-mile table, you can calculate the approximate cost of any toll call between the principal cities of the United States, whether the call is interstate or intrastate.

The second booklet, *Exchange Service Telephone Rates,* details the cost of flat-rate and message-rate business and residence telephone service in each major city. It is useful because sometimes a business may be using message-rate service without realizing that a flat-rate central office trunk may be available which would permit unlimited local calling without message unit charges.

In general, intrastate rates are *higher* per airline mile than interstate rates. Intrastate overtime charges may be substantially greater for both DDD and operator-handled calls, especially person calls. Rate periods differ, too. In 1973, only 12 states followed the interstate time periods.

Some states have call minimums of one or two minutes. Because

[2] *Long Distance Message Telephone Rates* and *Exchange Service Telephone Rates,* National Association of Regulatory Utility Commissioners, Box 684, Washington, D.C. 20044.

of these variations, any telecommunications executive with responsibility for controlling costs in states other than his own must familiarize himself with intrastate rate structures and peculiarities. Because of the higher rates applicable to intrastate calling, the cost-cutting techniques discussed earlier in this chapter will be even more helpful when applied to intrastate calls.

## SUGGESTIONS FOR MAKING DDD MORE EFFECTIVE

I hope you're convinced that you and your people should use the DDD network as much as possible. Now let's concentrate on making each call in the most effective way. Here are some ideas.

### OBTAINING AN UNKNOWN TELEPHONE NUMBER

To get an unknown telephone number, follow these steps: First check the letterhead or purchase order. If you still need the number, procure an outside trunk line. Dial the area code if the destination city is outside your area. Then dial these digits: 555-1212. The Information operator will give you your number; she may also give you a street address, if you need it. THERE IS NO CHARGE FOR THIS CALL. Do not ask your switchboard operators to secure such numbers. Dial them yourself or ask your secretary to do it.

### EXPEDITING YOUR CALL

To expedite your call, follow these steps:

1. *When the called telephone is answered:* Immediately say, "Long distance is calling Mr. X." This identifies your call as toll and expedites it to the extension.
2. *When the next person answers:* Introduce yourself fully so the call won't be "screened," thus taking more of your time. Say, "This is Mr. Jones at Acme Manufacturing in New York City. Do you expect Mr. Smith soon?" The question you asked gently forces the secretary to give you helpful information immediately rather than spending your valuable time on pleasantries.
3. *If your called person is telephoning on another line:* Courteously ask the secretary to put a note on his desk indicating that you're waiting on long distance and asking if he will be talking long.
4. *If he's talking on a lengthy call or if he's out of the office:* DON'T GIVE UP! Tell the secretary what you need. She knows

other available executives and the office filing system. Perhaps she can answer your question or connect you with an alternate person.

5. *If you can't get the needed information: Always* leave call-back instructions. Say, "Will you please ask Mr. Smith to call me back as soon as he returns?" Give your name once more and your extension number, area code, and telephone number. Most business people are becoming accustomed to returning station-to-station calls. Also, many of those you call have access to special low-cost toll facilities such as WATS, foreign exchange, tie lines, and even private dial networks. The secretary will have briefed him on the purpose of your call. He will have done the needed research and *will take less of your time* when he returns your call.

6. *For sensitive calls to customers:* Consider requesting that the customer call back *collect.* It has been found in most cases that the customer ignores the "collect" suggestion and calls prepaid. However, be sure to tell your switchboard operator that such an incoming collect call is acceptable.

7. *When telephoning one of your own employees:* If you are calling one of your own people and find that he is absent, and if you are using a low-cost facility such as WATS or foreign exchange, and if the called person does not have access to a similar facility, do *not* leave a call-back instruction. Find out when your man will return, give him a little extra time to be late, and then place a second call on the low-cost facility.

## CALL-SHORTENING METHODS

To save your valuable time and reduce company expenses, try these call-shortening methods:

1. *Organize your thinking!* If you jot down a short agenda on a scratch pad before picking up the telephone, you'll be sure to cover the ground thoroughly. Often, your called person will divert your train of thought with a question which is off the subject. Your agenda list will get you back on the track.

2. *Use the three-pronged introduction!* Most of us announce our name and company to the called person—and then we stop and wait for him to say something. He's a friendly person, and he won't talk about business for quite a while.

To gently force him to talk about what *you* want to talk about, add a leading sentence to your opening phrase. Say something like this:

> "Good morning, Mr. Smith. This is Ross Jones at Acme in New York City. I'm calling you about the quotation for oscillators you requested last week."

These sentences are delivered in a little package, without pause. Your man immediately starts talking about the oscillators, not about the weather or last Sunday's football game!

3. *When you've covered the agenda, say "good-bye."* Your called person has a loaded desk, too. He wants to get back to work, but he hesitates to be the first to terminate the conversation. Try using a summary phrase, such as:

> "Bill, I've certainly enjoyed talking to you this morning. I'm glad we settled all the details."

Now he knows that you think the conversation is finished, and you can easily terminate it without feeling uncomfortable. Note that the language is in the *past tense—enjoyed* and *settled.*

In dealing with a truly frustrating "long talker," you might use a more dramatic imaginary situation like this:

> "Just a minute, Bill . . . (aside) What did you say, Mary? Oh, Bill, could you excuse me? My secretary tells me that I have another long distance call waiting!"

## SUGGESTIONS ON EFFECTIVE USE OF "PERSON" CALLS

Should person-to-person calls be used at all? Very rarely. But all of us know certain people who are almost impossible to reach on the first attempt. They are talking on another line, away from their desk, or out of the office. Such "poor risks" may justify a person-to-person call. You may be able to cut the cost of that high-priced person call by following these procedures:

1. *If your man is reported out of the office:* The person responding to your call may tell the utility's operator when he will return. If you are doubtful about being able to talk with him at that time, you may tell the operator to cancel the call. Later, when you have returned yourself and feel that your man will also be at his desk, you will then place a DDD call.

2. *If your man is not available and an alternate person is suggested:* Consider carefully if the alternate person can give you the information you wish. If you talk with that individual, you will

still pay the person-to-person rate. If you think it advisable, you can cancel the call. Later, you can call your original person on a direct dialing basis, or you may place a DDD call to the alternate person suggested if your original person is not available.

3. *If you are calling one of your own people who is traveling:* It is possible that your man has not yet reached his destination. Your utility operator will "leave word." She leaves your name and number with the person taking the message. When he arrives your traveler must make a choice. If he returns the "Operator 6" call-back, he will not have to pay for the call if you are absent. However, if he does not ask for "Operator 6" and instead makes a station collect call and reaches you, he will save the difference between the person rate and the station collect rate, but he will run the risk of missing you if you are away from your desk.

## CONFERENCE CALLS

Your telco's operator can arrange a conference call for you whereby up to 10 different individuals in distant locations can talk together. The transmission level, which means clarity of conversation, is guaranteed to be satisfactory for a conference call of 10 persons or less. Conference call billing does not begin until all the conferees are on the line and ready to talk.

Special rates apply to conference calls, as they are not designated as person or operator-handled calls. The telco conference operator sets up the circuit in a way which will result in the least cost for your call. For instance, if you are in San Francisco and wish to talk to your managers in Boston and New York, you might expect to pay circuit charges from San Francisco to both points. Instead, your operator may set up the circuit from San Francisco to New York and then add another circuit between New York and Boston. The conference call rate will always be less than the sum of the standard long distance rates between the points involved.

If you are interested in making conference calls which involve up to four individuals in your office and one or two individuals in a *single* distant location, a special switchboard or console conference circuit can be installed by your telco. See the section "Switchboard Conference Circuits" in Chapter 11.

CHAPTER 2

# REDUCING THE COST OF LOCAL MESSAGE UNITS

A "message unit" is an accounting term which is used in connection with the cost of a telephone call made within a defined geographic area surrounding the location of your office telephone system. Message units are priced at so many cents per unit. A typical cost per unit in 1973 was 7 cents. Message unit charges apply only in certain geographic areas. In other areas, the user of telephone service rents a "flat-rate" trunk instead of a "message-rate" trunk. The flat-rate trunk is much more expensive but it permits unlimited calling to the local area without payment of per-call charges. Message-rate trunks may be priced from $5 to $12 monthly. In some telco territories, the trunk rental cost may include a certain number of message units. If additional calls are made beyond the fixed allowance, the subscriber is billed appropriately. In a few areas, the subscriber can choose between the flat-rate or message-rate trunk, using the trunk which will result in the lowest cost per local call. See the section "Selecting a Proper Trunk Configuration" in Chapter 10.

In some cities, a single message unit is charged for any call made within certain geographic limits, regardless of the length of the call. However, if the call is dialed to an area beyond the originating area, multiple message units may apply. For such calls, a charge of two to six message units may be made for the first three to five minutes of conversation. After the initial period, additional message units are charged for segments of overtime.

Automatic metering equipment tabulates the number of message units applicable to each local or suburban call. The number of units accumulated each month is shown on the telephone bill, with the dollar amount chargeable. Except in a very few cities, there is no message unit itemization on a per-call basis. Thus, it is difficult to analyze the lump-sum billing. Here are some suggestions on controlling this important element of telecommunications expense.

### TABULATE LOCAL CALLS

In a very small office with only a few trunk lines, the telephone user can keep a simple tally of local calls, assuming that a single message unit applies to each local call, regardless of length or distance. The tabulation is totaled at the end of the billing month and compared to telco's record of local calls. If a substantial discrepancy is noted, the telco business office is asked to check the meters. Special equipment can be attached to the meters to verify their accuracy.

### TRAIN EMPLOYEES TO USE THE TELEPHONE DIRECTORY

Telcos are filing new tariffs which provide that message unit charges will apply to requests for directory service from subscribers in the local area who have been furnished directories. Exceptions will be made for new numbers not found in the directory and for calls received from distant points via 555-1212, the universal information number. Management should distribute city telephone books to all users and should instruct them to look in the book before they dial the Information operator.

### LOCK TELEPHONE AFTER HOURS

Inexpensive key-operated locks are affixed to the dial, preventing any call dialing. After-hours personnel are then unable to use the office telephone system for local or long distance personal calling.

### RESTRICT TELEPHONES

The term "semirestricted," applied to an office dial telephone line, means that the user cannot dial a local call. Special wiring in the telephone equipment room produces a busy signal when the user dials the digit which normally would give him an outside line. In areas where single or multiple message units apply, the restriction of appropriate lines will produce a substantial saving in message unit and toll charges. A one-time wiring charge applies, and there is no monthly rental charge. The analyst should use caution, however. I suggest the following procedures.

### Discuss Problem with Industrial Relations Department

In major metropolitan areas, the employee may expect the privilege of dialing local and suburban calls. Thus, removal of this privilege may create a grievance which could cause difficulty for the industrial relations department.

### Discuss Problem with Department Heads

You may wish to send a letter to your department heads, explaining that message unit costs are mounting. Perhaps employee personal calls are loading your outgoing trunk lines, causing complaints from busy executives when they cannot secure an outside line. Ask the department head to carefully survey the job responsibilities of his people. In many companies, telephones have been restricted incorrectly. The department head should be given a list of presently restricted phones and asked to verify if the restriction should be continued. He should then be asked to list the names of employees with unrestricted phones whose job duties do not require dialing outside calls. If you are using any type of Centrex (see Chapter 9), ask the department head to list not only the employees who do not require outdialing but also those who should not be receiving calls from outside the Centrex system. This is because all Centrex tariffs offer a special "fully restricted" Centrex line which can be used to make or receive calls within the Centrex system, but which cannot make or receive outside calls. This type of Centrex line is usually available at much less cost than the unrestricted line.

### Write to the Affected Employees

When an outdialing privilege is removed, the employee must be told *why*. A carefully worded letter, stressing the need for ready access to trunks for company business, may soften the blow. The employee may be told that he can continue to dial important personal calls. He is told to see his supervisor and explain the reason for the call and is then directed to an unrestricted telephone where the call may be made under supervision.

### Instruct the Switchboard Operators

Those with restricted phones will quickly think of dialing the operator and requesting an outside line. Your operator should ask the identity of the person requesting the line. If the name is unfamiliar, the employee should be referred to his supervisor. Frequently an unrestricted caller may be in an area served only by a restricted phone. He must necessarily dial the operator to secure an outside line.

## FIND OUT WHERE CALLS ARE DIALED

In Chapter 4 I will describe how foreign exchange (FX) trunks help reduce message unit costs. However, before you can determine what FX trunks are needed, you must determine the exchanges to which your message unit calls are dialed. There are two methods of securing this information:

1. Ask your telco marketing representative for help. In some telcos, equipment can be attached to your outdial trunks which will record the exchange prefix and number on local calls. A charge may be made for this service and for the resulting tabular run. If possible, ask telco to arrange the printout so that calls are grouped first by exchange and then by telephone number within the exchange. By inspecting such a tabulation, and applying your knowledge of known customer and vendor telephone numbers which are called frequently, you may be able to spot flagrant use of your business lines for personal calls.
2. If no mechanical assistance is offered by telco, supply your users with a ruled pad showing the date, telephone number, and approximate call length. Use a separate piece of paper or tab card for each call. Accumulate the slips for a period which seems typical of business volume—perhaps a week or a month. Then sort them by exchange and within the exchange by telephone number.

Now, by either method, you will have accurate information for considering the use of FX trunks or special extended calling area services which may be offered by your telco.

## CONSIDER SPECIAL EXTENDED AREA SERVICES

Some telcos suggest that the subscriber pay a higher monthly rate per trunk for a service which permits toll-free or single message unit service over a wider geographic area than the normal calling area. Before subscribing to such a service, it is important that you study the quantity, frequency, and length of calls otherwise subject to multiple message unit charges.

### CONSIDER USE OF FX TRUNKS

You will learn in Chapter 4 about bringing a distant telephone number to your switchboard. When a user wishes to call that particular geographic area, he dials the operator and requests the special trunk. Or he can dial a code which connects him directly to the trunk.

Let us suppose your company is located in Manhattan. You have a group of customers located in Nassau County, Long Island. That area is within the New York City message unit area, and a charge of six message units is made for the first three minutes of each call. The frequent calls to these customers are adding substantially to your billing for local message units. You should ask your telco marketing representative to quote the cost of an FX trunk that comes to your switchboard from the closest exchange which would include the Nassau County area. You then rent one or more trunks from that Nassau County exchange. Instead of dialing "9" and the Nassau number, your user dials your operator and requests the Nassau trunk. If it is busy, the operator lists the caller's request and lets him know when the trunk is free. If your company needs two or morc lincs, you may decide to provide "dial access" to them. The user dials a special code. If a line is free, he hears a dial tone. He then dials the seven-digit Nassau number.

FX trunks are subject to a minimum usage charge of only 30 days. Thus they can be added or disconnected as needed, so it is easy to experiment with them. You may be sure that if there is a sufficient volume of calling, and if you can discipline your users to ask for or dial the special trunks, you will decrease your message unit cost substantially, even after allowing for the additional cost of the FX trunk. You will pay a monthly rental rate for the foreign trunk, plus a mileage charge.

### EDUCATE THE CALLER

In the example just described, there is an essential element: the user must remember to ask the operator for the Nassau trunk. To help employees remember how to use the telephone system properly, management must provide training as a part of the indoctrination procedure for new employees. Refresher training in the form of meetings and bulletins may help remind existing employees how to save time and also reduce cost for their company.

Let's consider another example. The central portion of Los Angeles contains offices of many financial institutions. The customers of those institutions are scattered around the Los Angeles area. Calls to many of them may be billed at multiple message units for the initial period, with additional message units for overtime.

A downtown bank installs some Long Beach FX trunks to reach customers in that area. To reach Long Beach and its surrounding areas, the user dials "8." Calls to other customers in the downtown area are made over local trunks by dialing "9." However, if an employee should incorrectly dial "8" and then a local downtown number, multiple message unit charges will apply. This is because the call is being received in the Long Beach exchange and automatically routed right back to downtown Los Angeles, where the call originated. This mistake not only increases message unit charges but occupies a trunk. If all Long Beach trunks are busy, some executive may be tempted to make a message unit call, further adding to message unit expense.

To correct this problem, management gives each user a booklet or a printed sheet of paper which can be taped to a desk slide. The employee is instructed to dial "8" for all the exchanges listed on the sheet, which are shown in both numeric and alphanumeric order. Management has carefully studied message unit charges and knows that use of the Long Beach trunks will reach those particular exchanges at least cost. If the exchange prefix wanted by the user does not appear on the special list, he knows that he should dial "9" for that call.

## SECURE ADEQUATE INFORMATION ON MESSAGE UNIT CHARGES

In some areas, the local utility's telephone directory gives complete information about the cost of message units, both for initial and for overtime periods. In other areas, management must request this information from its telco. Most telco marketing people are awake to these problems and will try to supply information on exchange areas, FX rates, and message unit costs compared with the cost of FX circuits and toll calls.

## CONSIDER USE OF SPECIAL CENTREX RESTRICTIONS

If you are a Centrex user, there are many ways in which restriction can be arranged to reduce message unit costs. For instance, a group of telephones may be semirestricted so that they cannot be used for dialing outside calls. Another group can be used to dial local calls which have unlimited talking time and are charged at a single message unit. A third group of phones can be used to dial any local or suburban call within the message unit area of the city. A fourth group can be used for dialing all message unit calls and toll calls. A fifth group consists of fully restricted ("interior") Centrex telephones. They cannot be used for receiving or making outside calls.

## CONSIDER USE OF DIVERTING EQUIPMENT

In some areas served by independent telcos, it is possible to rent special diverting equipment which will prevent an employee from making a call incorrectly. If he dials "8" for a number which is most economically reached by dialing "9," the equipment will refuse to accept his call. He may hear a recording which instructs him to dial correctly, or the call may be routed to the switchboard operator so that she may instruct him in dialing procedure.

As the interconnection of customer-owned equipment gains momentum, you may find that manufacturers of diverting equipment will quote you on your particular needs. Your telco will then be required to interconnect such equipment for you, possibly through interface devices which will carry a small monthly rental cost. This equipment is very useful to any organization located in major metropolitan areas. It eliminates employee error and cuts employee training time and expense.

## CONSIDER USE OF INTERCONNECTED TELEPHONE EQUIPMENT

The highly sophisticated telephone equipment available from independent manufacturers contains many more restriction options than the equipment normally supplied to the customer by the Bell System and the independent telcos. Further, the options are tailored to the need of each individual telephone user. The class of restriction is usually made effective by a simple plugging operation in the equipment cabinet. The type of restriction can be changed at will. The manufacturers will be glad to give you details on these features.

## CONSIDER USE OF A LOCAL PRIVATE LINE (LPL) OR AN OFF-PREMISE EXTENSION (OPX)

The LPL (see Chapter 5) is a tie line running between two switchboards, between two dial systems, or between the desks of two or more people. The LPL usually runs from building to building or from a main location to a subsidiary location in the same city. A monthly mileage charge applies to the LPL, plus inexpensive monthly termination charges for the connection to switchboards, switching equipment, or telephones. As the LPL is a private line with 24-hour, 7-day usage privileges, it is not subject to message unit charges. If you have two points which are relatively close together, with heavy calling traffic, the cost of the LPL may be less than the cost of the message units applicable if your calls are dialed through your city trunks.

Message unit charges can be eliminated when an OPX (see Chapter 5) carries calls from another point in your message unit area to your dial system.

CHAPTER THREE

# WIDE AREA TELECOMMUNICATIONS SERVICE

WATS is a wholesale telephone service which offers substantial savings in the cost of toll calling when you make many long distance calls. A WATS line is a long distance line or "trunk." You can rent a WATS line on a full-time basis or in measured form. A full-time line is available to you 24 hours a day, 7 days a week. Measured-time service, which is available for use 24 hours a day, includes 600 minutes of conversation time accumulated during the calendar month, for a flat monthly rental. Additional talking time is billed as overtime at an hourly rate, which is charged in tenths of hours. Only one conversation can take place at one time on one WATS trunk. If more than one conversation is required simultaneously, additional WATS trunks must be added.

When properly installed and used, WATS can reduce toll cost by as much as 70 percent, depending on quantities of conversation and the distance traveled by the call. However, if WATS is added to your service without your proper understanding of its use, your total cost for long distance service may go *up,* not down.

WATS has become a *stylish* service. One of your customers may say: "I'll call you back on our WATS line." Your understandable reaction may be: "Perhaps *we* ought to investigate WATS." You call telco's marketing department; the line is installed within a comparatively short time.

Why does the utility offer you a service which can decrease their toll revenue? The answer is apparent when you examine the total long distance expense of companies after they have added WATS to their service. In many cases, total expense has risen substantially. This comes about because of the "stimulation factor." Users are told that their organization now has WATS service and that they should use WATS for calls previously dialed on toll. The inference is plain: WATS must be less expensive. Many people think that their WATS calls are "free." Thus, traffic which may have moved through the mails is placed on long distance circuits. Unless careful controls are applied, personal long distance calls multiply. The organization quickly fills its WATS lines; impatient executives are unwilling to wait for a line and place their calls on toll. The result is predictable: higher costs.

Professionals who advise management about telecommunications think WATS is an *excellent* service. But their methodology of investigating and installing WATS differs substantially from that of their telco friends. This chapter will guide you in determining whether you should use WATS, and, if so, how you can get the best possible results from the lines you install.

## INTERSTATE WATS COVERAGE

Except in Michigan, where two-way WATS service was being tested late in 1973, WATS service is one-way, either outward from your office, or inward to your office so that others can call you toll-free. Interstate WATS service does *not* include the state in which your WATS call originates. If you wish to make calls within your state, intrastate WATS service is almost always available from your telco.

Interstate WATS is offered in "bands." Band coverage is arranged by states, radiating outward from your state. Band 1 is the least expensive; Band 6 covers the entire mainland United States except for the state where your call originates. Band coverage is cumulative. Thus, if you rent a Band 6 WATS line, you may also use it for calls made into Bands 1, 2, 3, 4, and 5. A Band 2 circuit accepts calls for Band 1; Band 3 also accepts calls to Bands 1 and 2; Band 4 accepts calls to Bands 1, 2, and 3; and Band 5 also accepts calls to Bands 1, 2, 3, and 4.

Each WATS band includes a certain number of area codes. You can call any telephone number within the area codes included in the band you rent. In theory, a given WATS band includes approximately the same number of telephone subscribers regardless of geographic location. Thus, rental of Band 4 in Chicago would give you access to about the same number of telephones as rental of Band 4 in New York City. Your telco will supply you with a colored map giving complete information on the area codes included in each band and the cost in your area for full-time or measured-time interstate WATS service in each band.

## COST OF INTERSTATE WATS

The rates for interstate WATS are set by AT&T and regulated by the FCC. A 2 percent rate increase was authorized early in 1973. Rates for a given band will differ substantially from state to state because of population concentration patterns and differences in geographic distances.

In general, the WATS rate is uniform for any city in a particular state. However, certain states have been divided, with different WATS rates applicable in each section. In 1973 the divided states were:

| | |
|---|---|
| California | North and South |
| Illinois | North and South |
| Michigan | North and South |
| New York | Northeast, Southeast, and West |
| Ohio | North and South |
| Pennsylvania | East and West |
| Texas | East, South, and West |

The monthly rates in effect in March 1973 for full-time WATS varied as shown in Table 1; those for measured-time WATS varied as indicated in Table 2.

**TABLE 1 Rates per Month for Full-time WATS**

| Band | Lowest rate | Originating state | Highest rate | Originating state |
|---|---|---|---|---|
| 1 | $ 510 | Connecticut<br>Delaware<br>District of Columbia<br>Maryland<br>Massachusetts<br>New Jersey<br>East Pennsylvania<br>Rhode Island | $1175 | South California<br>South Texas<br>Washington |
| 6 | $1685 | Iowa<br>Kansas<br>Minnesota<br>Nebraska<br>North Dakota<br>Oklahoma<br>South Dakota | $1990 | Connecticut<br>Maine<br>Massachusetts<br>New Hampshire<br>Rhode Island<br>Vermont |

**TABLE 2 Rates per Month for Measured-time WATS***

| Band | Lowest rate for first 600 minutes | Rate for each overtime hour | Highest rate for 600 minutes | Rate for each overtime hour |
|---|---|---|---|---|
| 1 | $135.00 | $10.20 | $225.00 | $16.80 |
| 6 | $295.00 | $22.00 | $325.00 | $24.50 |

* The states involved in the highest and lowest rates are the same as those listed in Table 1 for full-time WATS cost.

## COST OF INTRASTATE WATS

As telephone rates within a particular state are regulated by the state's public utility governing body, rates are not uniform. In 1973, Rhode Island did not offer any type of WATS service. California does not offer statewide full-time service. Some states offer 10 hours of calling as the basic package; others offer 15 hours.

Here are a few examples of WATS intrastate pricing at 1973 rates:

| | | |
|---|---|---|
| New York | Full time | $863.92 |
| | Measured time (900 minutes) | 318.99 |
| | Overtime per hour | 17.94 |
| Illinois | Full time | $625.00 |
| | Measured time (600 minutes) | 155.00 |
| | Overtime per hour | 11.50 |
| Texas | Full time | $815.00 |
| | Measured time (600 minutes) | 220.00 |
| | Overtime per hour | 20.00 |

## DETERMINING WHETHER YOU SHOULD BE USING WATS SERVICE

There is only one way to determine if WATS will save you money. You must divide your outgoing or incoming toll calls into WATS bands and then compare the cost of WATS for those bands with the toll cost. If you wish to do this work manually, a clerk must take a piece of accounting paper and put each toll call into a column for its corresponding WATS band. Your telco will give you the area codes applicable for each band, as maps are published for each WATS area.

However, this is very laborious work. A better way is to ask your telco marketing representative to arrange a WATS study for you. If you are interested in using interstate WATS, he will contact the telco toll accounting center which serves your area. Personnel there will program their computer to produce a WATS study for you during a future accounting period. Your telco marketing representative may work with the AT&T Long Lines department in securing your WATS study. The resulting computer printout will analyze all your long distance calls. A typical analysis will give you summarized information on the following calls:

Calls to points in your own state.
Calls to interstate WATS Bands 1, 2, 3, 4, 5, and 6.
Credit card calls.
Calls charged to your main number from some other number (third-number calls).
Calls placed beyond United States borders.
Incoming collect calls.

The information you will be given may vary, depending on the telco involved, but in general you will learn:

The number of calls made into each call classification as shown above.
The toll cost which you paid, in total, for each classification.
The number of minutes of chargeable time, in total, for each classification.
The number of 100 call-seconds (1 minute 40 seconds) for each classification. (These data are used by telco in their calculations for recommending the number of WATS bands you should use.)
The official recommendation of telco on the "best buy" if the figures indicate, in their judgment, that you should adopt WATS service. The "best buy" tells you which WATS lines you should install.

## ANALYSIS OF THE TOLL-CALL STUDY

If you make your own study on a manual basis, be sure to keep in mind that WATS lines permit only a single conversation in one direction at one time. How urgent are your calls? Can your users wait until a long-winded talker hangs up so the WATS line is available for the next call? Executives are usually an impatient group. They want to make a long distance call *now*. It will take some persuasion to get them to wait for WATS. But if they are willing to wait a reasonable time, WATS will bring you a handsome cost reduction.

Because there is always an overflow of long distance calls from WATS into toll, you must be very careful to allow for this cost in analyzing WATS recommendations, whether you are looking at your manual study or at telco's computer study. In general, we feel that you should be spending 125 percent of WATS cost in toll cost before you even begin to consider WATS, and 150 percent is a much safer figure. Let's suppose you are located in Southeast New York State. Your toll costs in Bands 1, 2, 3, and 4 total $2000. You are barely within the 125 percent range, as a Band 4 interstate WATS line costs $1580.

Here's another caution: In many telco WATS computer studies, the telco recommendation will provide sufficient WATS lines to ensure that 95 percent of WATS requests during the busy hour can be fulfilled within a 5- to 10-minute delay factor. In other words, when your user calls the operator during the busy hour, he either secures the WATS line immediately or he will be called back and given the line within a maximum of 10 minutes.

If you accept a recommendation based on such a short delay in the busiest time of the day, the WATS lines which telco will install are going to be less busy or even idle in the hours when normal calling traffic is lighter. You will be renting more WATS lines than you would need if you adopted a longer delay factor. Ask telco to calculate the number of WATS lines you would need for various delay factors, perhaps ranging from 10 minutes up to 30 minutes.

Further, you may find that telco recommends that you combine measured-time WATS service with full-time service. In this way, if the full-time lines are busy, your operator can connect your caller to the measured-time line. Such a recommendation may increase your costs substantially. This is because the cost per minute of measured WATS is frequently greater than the cost per minute of directly dialed, station-to-station toll calls. I'll be telling you about the characteristics of measured WATS later in this chapter.

Thus, don't immediately accept telco's recommendation on the WATS lines you should acquire. Look at your total traffic throughout the working day. Most business calls can wait for 30 minutes, even in the "busy hour." If you can sell such a delay factor to your callers, you will require fewer WATS lines, and your savings over toll will increase substantially. If you have urgent calls that cannot wait for WATS lines, your operator may connect your caller to an outgoing toll circuit on the switchboard, or your caller may be told to directly dial his call over toll circuits.

Be sure to test various combinations of WATS lines, keeping in mind that we recommend that aggregate toll charges in the WATS band you are testing must be at least 125 percent more than

WATS cost. While one Band 4 WATS line may seem to meet that figure, perhaps you could give your callers better service by arranging for two talking paths: a Band 2 line for Bands 1 and 2, and a Band 4 line used primarily for Bands 3 and 4. Contrast your toll costs in each band separately and in various cumulative combinations. Then order WATS in the configuration which *you* want, not what telco's computer recommends.

AT&T's computer study of outgoing toll calls is based on full minutes of talking. You will realize that the *final* minute of talking was charged to you as a full minute, although your conversation in that final minute might have been only a few seconds. Because of this factor, your actual talking minutes are somewhat less than the total talking minutes shown on the computer study. The difference does not justify your reducing the number of WATS lines needed to handle the required talking time, but you will have a small safety factor built into your calculations.

## Inward WATS

Before we leave your initial study procedure, let's talk about analyzing your need for *inward* WATS. Do people call you collect? Are they your *own* employees? Are they customers? Dealers? Distributors? Could you increase sales by using toll-free calling?

There are several methods of extending toll-free privileges to those calling you. In Chapter 1 we described the correct methods of calling collect and how to use "Zenith" and "Enterprise" inward toll calling. You should review that section now. Inward WATS gives you still another method of extending toll-free calling to the originator of the incoming call.

Inward WATS is *very* stylish. Your telco will be glad to talk about it. The tariff provides that you must install sufficient lines to give your callers a reasonable grade of service. It is very frustrating for your customers to call an "800" number and find it busy. Thus, your telco may recommend that you combine full-time inward WATS service with measured-time service. Or they may recommend *all* measured-time service.

Should you adopt their recommendation? First, make sure that the data they have used are correct. Your telco has the capability of determining *inward* demand for conversation with your main office location, providing you can furnish them with the telephone numbers of the firms or branch offices which call you. They may secure those toll statements and divide those toll calls into WATS bands. Then they will make a recommendation on the number of full-time and measured-time inward WATS lines which will be required. They will also analyze the *collect* calls which appear on your own statement, as well as the *third-number* calls which are recorded when one of your employees tells the distant operator to charge your main number for the call he is making.

It is possible that your telco may not have the manpower available to make the needed inward WATS study for you. You might ask your marketing representative if contact has been established with the local representative of AT&T's Long Lines department. Sometimes Long Lines can be very helpful in assisting the local telco staff. If such assistance is not available, you should develop the information in conjunction with your own staff. Ask your distant offices to send you their toll statements. Have one of your clerks divide the calls to your main telephone number into WATS bands and also analyze the collect and third-number calls which appear on your own toll statement. You will now have the cost of incoming long distance, collect, and third-number calls, as well as the number of calls. Depending on the accounting system of the telcos which serve you and your branches, you may also know the length of each long distance call in minutes. If you do not have that information, the various telco business offices can supply it to you. Once you have the total demand for conversation in minutes, the number of long distance calls, and their cost, you can determine the configuration of your incoming WATS system.

Should you now act upon telco's or your own information and install inward WATS service? Not until you have compared that cost with the cost of operator-handled calls on a station-to-station collect basis. Remember that a collect caller dials the utility's operator to make the collect request. The operator then dials the call over the long distance circuits available. It is possible that the call may be completed more quickly over those circuits than it will be if the caller tries to dial the "800" inward WATS number and finds it busy. Furthermore, the operator-handled rate may be less expensive per calling minute than the indicated rate per minute for the combination of inward WATS on a full-time and measured-time basis. An informed decision cannot be made until you do the needed calculations and then weigh the nondollar factors of possible increased speed for collect calling or the stylish appeal (everybody's doing it!) of the "800" inward WATS number.

If you *should* decide on inward WATS for use of your *own* people, perhaps including your dealer-distributor organization, be sure to read the next section.

## USING MEASURED-TIME WATS CORRECTLY

Regardless of whether you are considering measured-time WATS for inward or outward service, it is important to understand the steps which must be taken to decide if measured-time WATS is indeed your most economical choice. In other words, will it reduce costs? Here are some suggestions.

### Calculate the Average Call Length

If the current long distance toll statement received monthly with your telephone bill does not show the number of minutes of talking time for each call, you can secure this information from the stub end of your bill, which is retained in the business office of your telco. Some of the companies will lend you the entire packet of statements for a typical month, or they may be able to duplicate this information and send a photocopy to you. Sample this information adequately, or add the total minutes of talking time for toll calls in the typical month; then count the number of calls and divide these figures to get the average length of calls in minutes. You may wish to refine your analysis further by establishing separate call-length categories for outgoing calls, incoming collect calls, credit card calls, and third-number calls (see Chapter 1).

### Determine the Geographic Points to Which Your Calls Are Made

Your monthly toll statement will give you this information. If your statement shows the number of minutes talked, you can pick calls of one, two, or three minutes and assume that the cost shown is the standard toll rate to that city. You can now secure the cost per minute to the city by dividing that call cost by 3.

### Calculate the Per-Minute Cost of Measured WATS

Except in certain intrastate offerings, such as the California measured WATS service, the telcos sell measured WATS in a 600-minute package at a flat rate. For instance, a state may sell statewide WATS coverage on a measured-time 600-minute basis at $500. You can easily secure the cost per talking minute by dividing 600 minutes into $500. You now contrast the cost per minute on measured WATS to the cost per minute on toll. Multiply these minute costs by your average call length, expressed in minutes and tenths of minutes, and you will secure the actual cost of measured WATS service for comparison with your present toll cost. Subject to the exceptions mentioned in the next paragraph, you will find that diverting any toll call to the measured WATS circuit may increase your cost. In addition, instead of making unlimited toll circuits available to your callers, you will be restricting them to only the measured WATS circuits furnished by your telco.

### Use Your Calculations for Decision Making

If your average call length is two minutes or less, measured WATS may prove to be a real communications bargain. For instance, the trading departments of investment banking firms may be able to conclude a transaction in half a minute or less. If traffic does not justify the much higher cost of full-time WATS, measured WATS may appreciably reduce toll expense.

When you are talking over longer distances, and when traffic is well over 600 minutes a month but does not approach the theoretical full-time WATS capacity of almost 12,000 minutes monthly, measured WATS may reduce toll costs if sufficient overtime hours are used.

Once you have used the initial period of 600 minutes, each overtime hour is billed to you at approximately 20 to 25 percent less than the rate per hour for the initial 600-minute period. Consequently, once you achieve substantial volume in a particular measured-time WATS band, the weighted average of cost per minute for the basic period and for overtime may bring the composite figure down below your cost of talking on toll. Again, the average length of your conversation is critical and no conclusion can be drawn without such a calculation.

Here's an important suggestion: *Do not permit outward measured WATS lines to be used after 5* P.M. *or on weekends.* The sizable reduction in toll rates at off-peak hours makes the cost per minute of toll conversation far less than the cost per minute of measured WATS conversation, despite overtime discounts in the measured WATS tariff.

### Control the Problem of WATS Overflow to Toll

Your utility may emphasize that use of measured WATS gives you more talking paths during the busy hour when circuits are needed. Unless your use of measured WATS fits the exceptions mentioned above, you should disregard this marketing approach. Instead, your operator should be instructed to use a toll circuit for a WATS call if the originator cannot wait longer for a full-time WATS line. This will ensure the lowest DDD cost for the overflow calls.

## Consider Trunk Cost in Analysis of Measured WATS

If you decide that the cost per minute of measured WATS is more than the cost per minute of the typical toll call to destinations in a particular WATS band, there is one more calculation which should be performed before you make the decision to use toll. Your toll calls will require an outgoing trunk from your switchboard. In certain localities, these trunks provide unlimited calling without message unit charges. Such trunks are always more expensive than message unit trunks, and they may rent at upwards of $30 monthly in some localities. The cost of any needed additional trunks should be added to your economic study of toll calls versus measured WATS.

## Use Inward and Outward WATS to Satisfy Total Demand

Earlier in this chapter, we suggested that you ask your local telco to calculate the demand for inward toll calling from your customers and your own employees. If you impose sufficient discipline on the callers in your main office where your outgoing WATS lines are located, the following use of inward WATS will decrease your toll and WATS cost far beyond any telco recommendation.

First, calculate the *number* of calls which are to be made monthly on a toll-free basis. Next, itemize the information you need from each caller so that a prompt call-back could be made on your full-time *outward* WATS circuits from your main office. Such information usually includes:

1. Caller's name
2. Caller's area code, telephone number, and Private Branch Exchange (PBX) station number
3. Date and time of call
4. Name of main office person who should call back
5. Identifying information (order number, etc.)
6. Degree of urgency (how soon must the call-back be made?)

We suggest that a special two-part form be prepared on which the above information will be recorded by the answering clerk.

Now, assume that this information can be recorded manually within 30 seconds. Divide the figure for the total number of calls by 2—and you will know the number of *minutes* you will require for inward WATS. For instance, a single Band 6 circuit, covering the entire country, will permit recording of 1,200 call-back requests in the 600-minute basic period. If overtime is required, it is available, of course.

You order the needed line or lines. The line is terminated on the switchboard to ensure continuous coverage, but it is also terminated on a telephone located at a clerical position which is kept manned throughout the working day. The answering clerk is not permitted to converse with those calling the "800" number. She records the needed information and hangs up. She keeps the duplicate form copy and sends the original to the department involved in the call-back. The departmental manager must take responsibility for seeing that a call-back is made within the span of time required by the caller. It may be necessary to establish some kind of priority on outward WATS service if the caller is waiting at a coin station or other noncompany telephone for his telephoned call-back.

Let me emphasize that the system just described requires rigorous discipline to make sure that those entrusted with call-backs will indeed make them in sufficient time to satisfy your callers. If this is not possible, it may be better to use outward WATS to poll your own offices which produce constant inward toll. There will be wasted calls when there is no traffic inward-bound, but the polling method on full-time WATS may be superior to the use of expensive measured-time inward WATS circuits for lengthy conversations.

## CONNECTING WATS LINES TO YOUR TELEPHONE SYSTEM

Because a WATS line is a trunk, it can be connected to any telephone instrument directly, or it may be connected into your PBX dial equipment. From the equipment the WATS line can run to your switchboard, and through special switching equipment the operator can switch the circuit to a group of related telephones, to a single telephone, or to data processing equipment.

If your circuit is connected into your PBX dial equipment, you have two choices for securing access to the line. You may request it from the switchboard operator, or you may dial a special code on your telephone to secure the line. The latter method is called "dial access." We feel that control of WATS circuits by your switchboard staff is far superior to dial access. This is because the operator can accept requests in chronological order. If accounting requirements demand it, she can prepare a charge ticket for the call.

If the proper WATS line is busy, she may be willing to assign a vacant line in a higher band so that the call will not be delayed. If your organization is located in the East, and your operator has access to Band 6, she may utilize that band in the earlier morning hours for completion of calls to lower bands. This is because people in Mountain and Western time zones are not yet at their desks because of the time difference.

It should also be pointed out that dial-access WATS opens your lines to personal calls. We estimate that 10 to 15 percent of dial-access WATS usage is noncompany business. Further, executive time is wasted. A busy executive will not wait until busy signals clear—he dials his call on an expensive toll circuit.

To combat this time-wasting feature of dial-access WATS, certain telcos have proposed "Automatic Route and Advance" WATS dial-access systems, especially in Centrex installations. This use of WATS is *very* expensive. This is because the equipment first tests the correct WATS band when the caller dials the special WATS code for the proper band. If lines in that band are busy, the equipment tests the next *higher* band. If those lines are busy, the call is routed automatically to toll circuits. Thus, "Automatic Route and Advance" will tolerate no delay whatsoever. The cost of the circuits and uncontrolled toll calls will be far in excess of a properly designed, operator-controlled WATS system.

## SOME SUGGESTIONS ON THE USE OF OPERATOR-CONTROLLED WATS

Switchboards and consoles differ in their WATS operation characteristics. Your telco engineers must tell you the capabilities of the switchboard or console serving your business. In general, if you dial "0" on your PBX system and ask the operator for a WATS line, she can extend the WATS dial tone to you *on that connection* if she is operating a cord switchboard. If she is using a console, such direct WATS extending may not be possible. Instead, she may ask you to hang up so that she can call you back and extend the dial tone. In certain systems, when she calls back you may have to dial a dial-access code and then the city access code, area code, and number. Here are a few specific suggestions on WATS handling at the switchboard.

### Consider Operator-handling Time

Once you dial "0" and request a WATS line, your operator must take several actions. She talks to you and records the details of your call. She checks the availability of her WATS lines. Meantime she may be interrupted by the need to answer an incoming call, thus delaying her response to you. She then connects you to your line or says she will call you back when the line is available. The call-back takes a few more seconds. Consequently, operator-handling time must be part of your WATS economic study, as you must be sure your current switchboard staff can add WATS operations to its current workload.

The Bell System uses 100 call-seconds (1 minute 40 seconds) as a standard operator-handling time factor for WATS calculations. In general, we agree with that standard. To be conservative, you may wish to use a two-minute operator-handling time per call. While you may be reluctant to rent another switchboard position or hire an operator to handle WATS, we believe that WATS economy may easily justify this expense. Further, the added labor and switchboard rental cost should be more than offset by saving the cost of the personal calls placed on WATS lines when the user adopts dial-access WATS.

## Concentrate WATS Operations at Specified Positions

As in most office operations, specialization may offer you greater WATS operating efficiency. When several switchboard positions are operating, it is best to have one or two operators concentrating on speedy handling of WATS lines. Requests for WATS received by other operators are ticketed and passed down the board to the specialists. You may find it is best to channel all WATS requests to an off-switchboard position which is adjacent to the switchboards themselves. Perhaps there is a desk in the switchboard room which can be used. A six-button telephone is installed with two or three working lines arranged in hunting sequence. Users call a specific three- or four-digit number to make their WATS request.

It may be more convenient if you arrange for one-digit dialing for WATS requests. Ask your telco representative to connect a "level" of your dial system to the WATS request position. A "level" is any series of 100 numbers. For example, you do not have any telephone numbers starting with the digit "7." Thus, "level 7" may be available for single-digit dialing. The wiring can be done so that the calls can be answered at the switchboard itself or at a telephone instrument installed at some other location.

The answering clerk tickets the WATS request, time-stamps it, and delivers it to the appropriate switchboard position. We recommend that waiting tickets for each WATS line be filed chronologically in a small box. The WATS clerk can take the responsibility for filing the ticket in the box, thus making a further saving in time for the WATS switchboard operator.

### Eliminate Call Dialing by Operators

To speed WATS service, you should ask your callers to dial their own WATS calls. The operator extends the WATS dial tone and then leaves the line. To protect the lines from being loaded with personal calls, you may require that the caller give the name of the person being called as well as his telephone number when the original WATS reservation request is made.

### Use WATS After Hours

While I have spoken strongly about the disadvantages of dial-access WATS, I certainly favor its use after the switchboard has closed. An inexpensive switch is mounted at the switchboard. The operator turns the switch when she closes the board. Now any telephone user can secure a WATS line by dialing an appropriate code. If the management wishes to restrict WATS usage after hours, the lines can be connected on a night-line basis to certain designated telephones located in places announced to authorized night users. WATS lines may be used after hours by data processing departments for sending data from the computer to branches or for polling branches for incoming data to the computer. The necessary interfacing equipment can be rented from telco or purchased from outside manufacturers and interconnected to the WATS line or lines. When a WATS line is to be used for sending data after hours, it is wise to install the needed switch in the data processing department rather than at the switchboard. This will prevent accidental disconnection of data transmission with a corresponding loss of data.

The off-premise extension (OPX; see Chapter 5) may help ensure that you secure full value from WATS. A regular station line is run from your switchboard to the residence of your sales manager. After hours, your operator converts the WATS lines to dial access, or she may connect a WATS line directly to the sales manager's three-digit number which rings in his home. If the former, he dials the WATS access code for the appropriate line. If the latter, the dial tone he hears is the tone of the WATS line directly connected to his OPX. The cost of the OPX is the regular PBX station charge, plus the mileage charge from your telco central office to the central office which serves his home.

## Be Careful about "Scheduled Calling"

Some companies provide a special switch at the switchboard which will remove a WATS line from operator control and deliver it to a specified telephone or group of telephones during regular working hours. That particular department or individual presumably makes WATS calls constantly during the time the line is assigned. Such WATS usage tends to stimulate personal calling when business calls have been exhausted. Also, the individuals may not have prepared their calling list, so the line sits idle. Wiring can be arranged so that the switchboard operator can determine whether or not the line is being used. If not, she can reclaim the line or report the facts to management.

As mentioned earlier, you may be asked by your branches to poll them at scheduled times for incoming voice or data traffic. This eliminates their making toll calls. They organize their business and hopefully are ready to talk to home office personnel when your operator calls at the appointed time. These arrangements may interfere with the proper scheduling of outgoing WATS calls, resulting in excessive delay. Executives make toll calls instead, increasing cost. We think it is better to use inward WATS lines for taking call-back requests from branch offices. Calls on outward full-time WATS can then be arranged at specific times. If your company has a Telex network (see Chapter 14), Telex provides an excellent device for sending a short message: "Joe Brown, please telephone me at 2:00 P.M. this afternoon." Such a message can easily be sent for pennies within the tenth-minute Telex minimum.

## Be Cautious about Using WATS for Band 1

In many cities, a WATS user may be quite close to the state boundary. For instance, Parkersburg, West Virginia, is within a few miles of Marietta, Ohio. A Parkersburg user of Band 1 WATS service may find that a toll call to Marietta will cost less than his actual cost for a full-time WATS line call of similar length to Marietta.

Further, if a WATS user dials a Band 1 call on a Band 4 line, he may force a Band 4 caller to use a toll circuit because the WATS line is busy. You should calculate your actual cost per minute of full-time WATS service, using minute-usage figures furnished by telco. Then calculate the toll cost for your average-length call to that same city. You may wish to furnish a map to your WATS users with a circle drawn on it. Points within the circle should be dialed on toll circuits because they are less expensive, or perhaps only slightly more expensive, than WATS calling.

### Experiment with "Patching" Calls from Outside

If you are traveling and you wish to call your office so that you can make a long distance call over your company's outward WATS line, it may be possible for your switchboard operator to connect your incoming trunk call to one of the WATS lines which terminates on the switchboard. A wiring change may be required to do this. Your operator can experiment, and if there is difficulty you can ask telco for information on the needed modifications. As the WATS tariff clearly provides that WATS is to be used only for business purposes, your operator must be careful to connect only calls which are required for company business. The operator can ask the caller for the telephone number she is about to dial and also for the name of the organization he is calling.

The "bridging" of an outside trunk call to a WATS line, tie line, FX line, or dedicated network may result in transmission difficulties and impair the clarity of conversation. This type of connection usually would not be appropriate for any type of data transmission.

### Reduce Trunk Lines if You Install WATS

Remember that a WATS line is an outward trunk. If you install WATS, your people will be making less outdial calls on your outgoing central office trunks. Thus, you will need less trunks. Once WATS is installed and operating successfully, you may be able to disconnect a number of your outdial trunks, thus making a worthwhile cost reduction.

### Remember Federal Taxes in Making Economic Studies

As a WATS line is a trunk, it is subject to taxes. All WATS lines are subject to the federal excise tax. Intrastate WATS lines may be subject to state and municipal taxes.

## CONTROLLING THE USE OF WATS

Once you install WATS lines, you will notice an increase in the total number of calls placed on long distance circuits, both WATS and toll. WATS has been widely publicized. Many telephone users look on their company's WATS service as "free." Consequently, the telephone is used more and more for nonurgent business which might be transacted by Telex, TWX, private wire networks, or the United States mail. Consequently, management must work out defined WATS controls which are enforced. Here are a few ideas.

## Guarantee Call Completion Time Interval

In an earlier paragraph I suggested that you attempt to sell a 30-minute delay factor to your users. That delay will become more acceptable if you will *guarantee* that any WATS call will be given access to a line within a specified period. The operator tickets the call, recording the time of the request. If a line isn't available at the end of the agreed waiting period, the operator calls the user to ask if he can wait longer. If he cannot, she gives him first priority on the next available line, or she connects the call to an outgoing trunk so that the caller hears a trunk dial tone and dials his call over the DDD network. We suggest that such WATS bypass toll calls be placed on switchboard trunks which are specifically assigned to WATS bands and are used for no other purpose than outgoing overflow calls from WATS to toll. Arrangements are made with telco to bill these trunks separately from the main telephone bill. At the end of each month management immediately knows the amount of overflow traffic. WATS lines can then be added or subtracted to maximize cost reduction efforts.

## Identify and Control "Long Talkers"

Every switchboard operator can tell management the names of callers who monopolize WATS, FX (see Chapter 4), and tie lines (see Chapter 5). We think such individuals should be asked to use toll circuits for their lengthy calls. The caller or the operator makes up an individual toll ticket for each call so that the accounting department can segregate that call and charge the appropriate budget. Further, we suggest that the operator be given authority to automatically extend toll circuits to such long talkers rather than connect them to WATS lines, ticketing the call appropriately. In announcing WATS usage rules to callers, specify a maximum conversation time. When the operator has time to observe the length of the call, ask her to make a daily log of those who exceed the maximum talking time. Management can then discusss the situation with the supervisors of the long talkers, asking for their cooperation in reducing the length of the calls.

## Consider the Use of Interconnected Devices for Limiting Call Length

A number of manufacturers sell or rent devices which will put a "tone" on WATS conversations after a preset number of minutes have elapsed. Some devices gradually increase the volume of the tone until conversation becomes difficult. Such devices may help materially in reducing the length of WATS calls.

## Audit Toll Calls Made to WATS Areas

Your toll statement from telco should be analyzed periodically to determine how many toll calls are being made to points which could have been reached by WATS. Telco will identify the telephone subscriber called if you do not recognize the number. Special forms are provided on which you can list the calls to be identified. The business office fills in all available information. You can then try to identify the department or individual who has not yet learned to use WATS. Educational work can then be done.

## Secure Periodic Statistical Information from Telco

Because WATS is a wholesale telephone service, your telco is not required to give you *any* statistics about the use of your WATS lines once they are installed. However, recording equipment in the telco rate centers can be programmed to give your telco marketing representative whatever information he may want about your use of your WATS system. Here is a partial list of the information which may be provided by the telco accounting office to your telco marketing representative:

1. Total minutes of talking time per WATS line for a typical month
2. Number of calls placed on each WATS line
3. Cost of those calls had they been placed on toll circuits
4. Computer printout of every call placed on WATS, in chronological order, showing
    *a.* Date
    *b.* Area code, exchange prefix, and telephone number
    *c.* Minutes of talking time
    *d.* Time of day call was placed
5. Computer printout of every call placed on WATS, arranged by
    *a.* Area code
    *b.* Exchange prefix within that city
    *c.* Telephone number within that exchange
    *d.* Date, minutes of talking time, and time of day call was placed
6. Computer printout which summarizes the number of calls and total talking minutes in each WATS band of a multiband circuit. If Band 4 is to be used primarily for calls to Bands 3 and 4, such a printout helps management correct WATS operating procedures, as management can analyze why Band 4 is being assigned by operators to service requests for Bands 1 and 2.

Any WATS user can readily perceive the value of periodic information of this kind. Most telco marketing executives are very cooperative in helping you obtain part or all of this information so that you can better analyze your use of WATS. In some telcos, this kind of information is furnished twice a year on a no-charge basis. Monthly figures can be secured from some telcos on the total talking minutes and the number of calls placed on each WATS circuit. In other companies, a charge is made for producing specific studies requested by the WATS user. The well-managed telco wants its customer to get the best possible value from special telephone services. Talk to your marketing representative about your needs and ask for his cooperation, but bear in mind that the WATS tariff does not *require* telco to give you any usage information whatsoever.

### Interpreting Telco's Information

In a nine-hour working day, and in 22 working days during a month, a WATS user has 11,880 talking minutes available on a single WATS line. If the line is operator-controlled, it will stand idle while the operator is recording information or when she is busy answering incoming calls and doesn't notice that the WATS line is free. Call-dialing time also must be subtracted from theoretical capacity figures. Consequently, we believe that 70 percent usage of the line is a reasonable efficiency factor. Seventy percent of 11,880 minutes is 8,316 minutes. You should check usage of your lines to see how close you come to that figure. If you are securing better usage, you have excellent operators. Or perhaps your data people are using your lines after the switchboard closes. If your usage is substantially less, and you have toll calls bypassing WATS, you should be doing some educational work on the cost-reduction advantages of "Wait for WATS," a slogan which you may wish to publicize.

The detailed computer runs listing every call are very helpful to management. You can spot numbers repeated day after day that cannot be identified as your own branch offices, customers, or vendors. A little checking with telco may reveal multiple personal calls to the same number. The computer study that groups all calls to the same number in one place is especially helpful in identifying personal calls.

In talking with your telco marketing people about the availability of statistical information on WATS, remind them that you recognize that the use of WATS for personal calls is prohibited by the WATS tariff. Tell them that their assistance in providing the computer studies will help you ensure that their tariff is being observed by your organization.

Remember that WATS is rented on only a 30-day minimum usage basis. Installation lead times are relatively short in most areas. Lines can be disconnected as your needs change. Thus, the alert administrator will constantly evaluate his "mix" of WATS lines. Suppose that he is using lines in Band 4, but routing calls to Bands 5 and 6 over toll lines. He asks his clerk to run a monthly total of the toll cost to Bands 5 and 6. When that cost starts to climb, perhaps one of the Band 4 lines should be changed to Band 6. For instance, in March 1973 a Band 4 WATS line cost $1375 monthly in Chicago. A Band 6 line cost $1735. Calls made on toll to points in Bands 5 and 6 totaled $550. But for an additional $360 a month these calls could be placed on the Band 6 WATS circuit with an excellent possibility of saving money even if some calls continued to overflow into toll.

As there is a three-hour time difference between the East and West Coasts, an East Coast user of a Band 6 line can talk to Western points effectively only during the six hours from 11 A.M. to 5 P.M., Eastern Standard Time. This means that the Band 6 WATS line has only a 66 percent loading during the Eastern working day if it is restricted to Western calls. As Band 6 includes all other bands, users can route overflow traffic on other bands to Band 6 until the West Coast offices open.

Constant experimenting with your WATS configuration will pay you handsome dividends. If you have adopted outgoing measured WATS because your conversations are less than two minutes in length, on the average, remember to check your overtime charges on your monthly WATS bill. In some companies, overtime on a single measured WATS service, plus the cost of the first 600 minutes, is more than the company would have paid for a full-time line. Telco's computers are not programmed to call your attention to such a situation. You must make your own calculation and change your lines accordingly.

If you are using more than one measured WATS line in the *same* band, and if the several lines are installed on your switchboard or on a telephone instrument, or are available by dial access, be sure to insist that telco bill those lines as a *group*. This ensures that telco will credit you with 600 minutes of prepaid rental on each line, even if one of the several lines has been used for only a portion of the 600-minute basic rental period. In any multiple line measured-time WATS installations, the first line in the sequence will accumulate the most overtime. It is important that unused segments of the 600-minute initial contract period be offset against the overtime before you are charged for overtime minutes. Group billing will help ensure that this is done. If you are using a separate WATS installation in a particular department, telco views that installation as an accounting entity, and any unused portion of the basic period cannot be credited against a separate WATS installation. If possible, secure figures from telco on the number of talking minutes charged against *each* measured WATS line to make sure that your WATS overtime bill has been calculated correctly.

### Evaluating Grade of Service on Inward WATS

At the time this chapter was written, AT&T had not perfected a method of tabulating total minutes of talking time and the number of calls made to inward WATS circuits. This is because they originate from many points, not a single point. Thus, the statistics you need to measure the service provided by inward WATS are best developed by your switchboard operators or by the clerk who takes the messages sent on inward WATS for call-back on outward WATS. Your operators can tabulate the number of calls received and, if their time permits, the duration of each call. In turn, telco can provide you with a "busy study" on each WATS incoming trunk or sequential series of trunks so that you can evaluate the possibility of adding additional trunks. The inward WATS tariff gives telco the right to insist that you provide a proper number of WATS inward trunks to minimize busy signals. If you are unwilling to spend that extra money, the tariff permits telco to remove your inward WATS service.

## WATS CONTROL EQUIPMENT MANUFACTURED BY INTERCONNECT COMPANIES

Several types of equipment are designed to help the WATS user reduce his costs and better use these expensive circuits. Because "interconnection" is still new, many vendors are just beginning to realize that there is a substantial market for this equipment. Here are some general comments and descriptions.

### WATS Usage Measuring Equipment

Several companies manufacture equipment which will count minutes and the number of calls on a specific WATS line. A pen recorder furnished by one company as an optional accessory measures the length of a particular call. Some of the equipment is actually interconnected through wiring to the WATS circuits. The use of such equipment may require an interface device rented from your local telco. Other equipment uses sensors that can be taped near the switchboard lamp, which glows when a WATS lines is being used. The sensor starts the measuring equipment when the light glows; it stops when the light is extinguished. As no wiring is needed into the circuit, the device is not subject to telco interface requirements. However, as you have learned in the preceding paragraphs, you may be able to secure regular reports on the number of WATS calls and talking minutes if you have good rapport with your local telco marketing representative.

### Equipment for Using WATS from Off-Premise Locations

Several firms are marketing a device which a traveling person can carry in his briefcase for use in connecting to his company's WATS lines. The unit is held to the transmitter of a telephone or coin station. It authorizes a special machine connected to the WATS line to "patch" the call from the distant telephone into the WATS line. The calling person then hears a dial tone, which is a WATS dial tone. He dials his telephone number.

Such equipment requires that the using organization install dial-access WATS rather than operator-controlled WATS. You have learned the disadvantages of dial-access WATS in previous paragraphs. However, if you are located in the eastern section of the country, your employees in time zones west of you may be equipped with these units, so they can access your WATS lines after your operator closes the switchboard and converts the WATS lines to dial access. The equipment is quite expensive; you will wish to weigh its cost against its utility in your particular WATS usage.

As this book was being completed, a major independent telephone utility announced a new unit called "Comex." In its intitial development, Comex could be used at any telephone equipped with Touch-Tone service. The device would permit the call to be connected with any type of telephone circuit available at a distant point. This concept may prove quite useful to organizations which have many dedicated facilities which could reduce toll cost if used by authorized persons located in other cities.

### Equipment for Automatic Queuing and Toll Ticketing

Many manufacturers are developing equipment which provides a printout identifying the particular PBX station number which originates a long distance call. The equipment usually shows an identification of the circuit used, the length of the call, the date, and other needed information. Using standard cost figures for calls of stated length and distance, the user can develop his own computer program to convert the information on the printout into charge-back information for accounting purposes. Such programming can also give the WATS user immediate information on WATS usage in each band, if desired. This helps the user change his WATS configuration as his usage pattern changes.

Several manufacturers are designing still more complicated equipment which will provide accounting on an individual basis. Each authorized long distance caller is assigned an identification code. That code is dialed into the control apparatus, plus the telephone number desired. The machine ranks the call in order of its receipt and selects the most economical channel for routing the call. It may choose between WATS, tie lines, FX trunks, and toll circuits. The machine then automatically dials the call, relieving the switchboard operator of WATS call handling, yet retaining the advantages of queuing and controlled WATS operation. One manufacturer provides cathode-ray tubes (CRT screens) at the switchboard, showing which call is next in line. Then the operator dials the call without the need for further paperwork.

Organizations which specialize in computer technology or which use computers extensively may offer special analytical service to those contemplating WATS usage or reappraising their use of WATS. These programs are quite expensive. Once the initial study is made, the offering organization normally will suggest a continuing progam for regular examination of WATS usage. We believe that the many analytical techniques mentioned in this chapter should help you use WATS effectively without the need for computer studies made by outside firms.

We expect that WATS will be offered by the telephone utilities for many years to come. The user should constantly survey the availability of peripheral control equipment such as that described above.

# CHAPTER FOUR

# FOREIGN EXCHANGE SERVICE

The term "foreign exchange" (FX) means that you are using a trunk which is brought to your telephone or to your switchboard and dial equipment from an exchange that is "foreign" or distant from your own telephone exchange. The FX line may come from another part of your city or from across the country. These trunks are rented at the rate in effect in their home community. In addition, a monthly mileage charge is added for the distance between the FX and your exchange. If the FX line comes from a point outside your telco's serving area or from a point outside your state, your mileage invoice may come either from your local telco or from the Long Lines department of AT&T. The telco which serves the area from which your FX number comes may bill you for the "local service" charge covering the FX number. The cost of terminating the FX line in your Private Automatic Branch Exchange (PABX) equipment will be billed as a part of the "local service" charge on your monthly telephone bill.

If you are interested in learning more about FX lines, your telco marketing representative will correlate the procuring of information and will also bring AT&T Long Lines representatives into the discussions if their advice is needed.

## USE OF FX CIRCUITS

Your FX line is a telephone number in the community at the distant end of the line. When you are connected to the FX line by your switchboard operator, or by dialing a code within your dial system, or by pressing a button on your telephone, the dial tone you hear is not a local dial tone but the dial tone of the distant telephone system. Now you can dial any number which is accessible from that exchange. If the distant community permits unlimited local calling on the trunk you rent, there is no charge for your local calls other than the trunk charge, the mileage charges to your office, and local terminal charges at your switchboard. If message unit charges apply at the distant location, you will be billed at whatever rate is in effect. However, you will not be paying a toll charge or a multiple message unit charge to reach the distant city. If you have numerous calls to many different telephones in the distant area, an FX line may save you a considerable sum. This chapter deals chiefly with the use of FX lines for reducing toll expense. Chapter 2 presented information on using FX lines to reduce local message unit expense.

## CONTROLLING THE CIRCUIT

An FX line is a "dedicated" facility. This means that you have full access to the service on a 24-hour, 7-day basis. FX lines are used by many companies for transmitting data after the office closes. In general, they are used very much like WATS lines. Consequently, apply your WATS line control system to your FX lines, too. Place them under operator control if they travel over distances beyond your local calling area. This ensures that users are taken in turn, and you are protected against the use of FX lines for extensive personal calls.

## USING FX TO RELAY TOLL CALLS

Suppose that you are located in New York City. Your company has a great many customers in the Chicago area, so your toll bills have been running very high. You have determined that a WATS circuit would be quite expensive, and you do not have to make many calls to other parts of the WATS area which includes Chicago. At the same time, you will have *some* toll calls to midwestern points such as Milwaukee. You decide to bring a Chicago telephone number to your New York switchboard. If you wish, this FX line can be arranged to give you access to the DDD network in Chicago. To reach Milwaukee via the Chicago FX line, you dial the area code and telephone number as soon as you hear the dial tone, as this is a *Chicago* dial tone rather than a *New York* dial tone. As the toll call originated in Chicago, you pay only for the cost of the call between Chicago and Milwaukee. At 1973 rates, that three-minute call would have cost 60 cents. In comparison, had you dialed the call directly from New York, you would have paid $1.15. That saving of 48 percent would look good to any financial executive!

You may wish to use your FX line exclusively for local calls in the distant area, realizing that if you use your FX line for relaying toll calls you will make it less available for local calls. Telco can wire your FX line so that it is "toll-diverted." If an area code is dialed, a busy signal is heard. In some areas, you can arrange the circuit so that it can reach multiple message unit suburban points, but toll calls cannot be dialed.

## ROUTING INTRASTATE TOLL CALLS OVER FX

There are important cost differences in the rate per mile for long distance conversations within a state and the rate per mile for a similar conversation which travels to another state. In almost every case, intrastate long distance rates are higher. I have just described a method of saving money by relaying long distance calls over your FX line. Now let me caution you about using an FX line for relaying calls from an interstate geographic location over an FX line which comes from the same state in which your call destination city is located. An example should help illustrate the principle.

Suppose that you are located in Newark, New Jersey. You call firms in New York City constantly, so you have rented several FX lines which bring New York City telephone numbers to your switchboard and PABX equipment. You have branch offices in Albany, Syracuse, and Buffalo. It seems logical that when you wish to place long distance calls to New York State points, you should use your New York City FX lines. This should not be done, because New York State long distance rates are substantially higher per airline mile than the interstate rate schedule. In general, always place your call on an interstate directly dialed long distance line if you are located in a state other than the one where your long distance call will terminate. Be sure to instruct your callers on correct dialing procedures. As mentioned earlier, your FX line can be "toll-diverted" to prevent its being used incorrectly for long distance dialing.

Your local telco can provide you with copies of the intrastate and interstate long distance rate schedules, which show the "rate steps" based on airline mileage. Or you can obtain the booklet *Long Distance Message Telephone Rates,*[1] which is published annually.

If an FX line is located within a state, the intrastate mileage rates will apply. If the line crosses a state border, the lower interstate mileage rate will apply, unless a special tariff is applicable. Such special tariffs are sometimes established between major cities which are very close together but separated by a state line. Your telco representative can give you full information on the costs which will be associated with your FX circuit.

## FX DIRECTORY LISTINGS

When you rent your FX number, you are entitled to both white page and yellow page directory listings. Thus, your customers and vendors in Chicago can look up your company name and find your number. When they dial that number, your FX line rings in New York City. You are providing toll-free service to Chicago customers so they may place orders easily. However, if your customer is using your line, your own sales or purchasing departments are blocked from making outgoing calls to Chicago. You have a choice between listing your number or keeping it unlisted. If you choose the latter alternative, you can release the number to selected customers, or you may wish to use the line exclusively for outgoing calls. If you are using several FX circuits which come from the same distant community, you may wish to arrange some of them in rotary sequence so that your distant customer needs to remember only a single number. If that line is busy, the equipment automatically switches the call to the next available line.

If you wish, it may be possible to arrange the installation of an FX circuit so that it operates on a one-way basis. The caller in the distant area has complete access to the line because you do not permit your operator to connect outgoing calls to the line. Likewise, you do not install the line on a dial-access basis. It appears only on the switchboard. Similarly, you can use your FX line only for outgoing calls from your switchboard or from your dial-access switching equipment. The FX number is not released to others in the distant city, nor is it listed in the city's telephone directory.

[1] *Long Distance Message Telephone Rates,* National Association of Regulatory Utility Commissioners, Box 684, Washington, D.C. 20044.

In some telcos, there may be a way of mechanically restricting an FX line to one-way operation, just as you can rent a one-way out-dial trunk or a one-way incoming trunk. As you can see, FX service is flexible, it reduces cost, and it may be a valuable marketing tool.

If you are considering the use of FX service within your own city, you may find it useful in reducing the cost of message units. See Chapter 2 for suggestions on how to arrange such circuits.

## CHOOSING THE PROPER CENTRAL OFFICE EXCHANGE FOR FX SERVICE

Each telco central office has a defined calling area where local calls can be made on an unlimited basis, or in message unit areas, for a single or multiple message unit charge. Most single message unit areas permit unlimited talking time, although tariffs are being filed to place time limitations on such calls. When you are bringing a distant telephone number to your location, be sure to define your needs.

Determine whether the calls you wish to make are spread evenly throughout the distant area or whether they are concentrated in a particular area. Then ask your telco marketing executive to select a telephone number in the distant city which will give you the maximum geographic coverage you need at minimum cost, particularly if your distant area is subject to multiple message unit charges from one central office exchange to another. You will be paying a mileage charge from your central office to the central office serving the telephone number you select. Try to make that mileage charge as low as possible by informing yourself about the calling area of the selected distant exchange. Can you reach most of the telephone numbers you call at only a single message unit charge?

Perhaps an example will help you to understand the planning procedure. Many FX trunks run between the cities of Tacoma and Seattle, because both cities have large business communities. However, rather than bringing a downtown Seattle number to Tacoma, organizations discovered that they could rent an FX trunk from an exchange serving the town of Des Moines, Washington. The Des Moines exchange is given toll-free calling into Seattle. As Des Moines is closer to Tacoma than Seattle, the mileage charges are less.

CHAPTER FIVE

# TIE LINES, OFF-PREMISE EXTENSIONS, AND LOCAL PRIVATE LINES

Users of long distance service should always consider the several alternatives which can be used in attaining their cost-reduction objectives. Three alternatives have already been covered: toll calls in Chapter 1, WATS in Chapter 3, and FX in Chapter 4. Three more options will be developed in this chapter: tie lines, off-premise extensions, and local private lines. Each of these specialized services is available from all telcos on both an interstate and an intrastate basis. These services are also available between the United States, Canada, Mexico, Alaska, and many foreign points.

A tie line is a circuit which connects two or more telephone systems. The systems may be PBX (a private telephone system operated by a manual switchboard), PABX (a private automatic telephone system which uses dial switching), or a combination of the two.

An off-premise extension (OPX) is a station of your telephone system. In Bell System terminology, the correct term for this facility is off-premise station (OPS). An OPX is actually a bridged instrument associated with an OPS. However, because OPX is the term commonly used in the literature, I am employing it in this chapter. The OPX is located at a point which is in a different location from the one which houses your manual or dial switchboard.

That point may be in an adjacent building, or it may be in the residence of an executive, or it may be situated in an office across the country. The distant user lifts his telephone handset and hears a dial tone. The tone is originating in the dial equipment at the headquarters location. The distant user can dial any other telephone in that system. He can dial "9" and secure an outside line in the distant headquarters city, unless his telephone is "restricted." If WATS, FX, or tie lines are available to the user via dial access, he can connect himself with those lines by dialing the appropriate code. If those lines are operator-controlled, he can dial "0" and ask the operator for the wanted circuit.

A local private line (LPL) usually serves two or more telephone *instruments*. For instance, a stockbroker may have a need for an immediate connection to his trader on the floor below. An LPL is arranged so he can depress a button on his telephone instrument, lift the receiver, and be immediately connected to the trader's telephone instrument. Usually an automatic ringing arrangement is included.

In general, these special services are billed to you on a mileage rate per quarter-mile or full mile. There may be other rental charges for switching equipment and for devices which connect the lines to your switchboard or dial system.

The mileage rates which were applicable in 1973 to interstate full-time tie lines are shown later in Table 1. So that you can visualize the approximate mileage charges you may pay for a voice-grade tie line between major cities, here are some figures showing approximate *mileage* charges. The small additional cost of connecting the line to your switching equipment is not included.

| | |
|---|---|
| New York–Los Angeles | $2200 |
| New York–Washington | $ 400 |
| Chicago–Dallas | $ 950 |

You can calculate your own estimate of mileage cost by finding the airline mileage between the cities and multiplying it by the figures shown in Table 1. Airline mileage information is shown in the guides listing airline schedules.

## TIE LINES

A tie line is useful when there are *many* conversations or quantities of data to be sent to a specific distant point or group of related telephones. Suppose your headquarters office is in Chicago and you have important branches in St. Louis, New York, and Dallas. You are constantly calling your company personnel in those cities. Conversations are frequent; many of them are lengthy. You do not need WATS, which would reach telephones throughout the East Coast: you want chiefly to reach the telephones in your New York office. A tie line from Chicago to New York may give you a handsome cost reduction from the toll charges otherwise applicable to those conversations. Similarly, you may be able to arrange another tie line which runs from Chicago to St. Louis and then on to Dallas. Your operator can route calls by using a ringing code, or your tie line may be arranged so that you can dial a code to reach a specific destination.

## INTERSTATE TIE LINES

Interstate tie lines may be provided by the local utility, or the lines may be arranged by the Long Lines department of AT&T, working through the local utility. Such tie lines provide "dedicated" service. This means that the line you rent is dedicated to your use on a 7-day, 24-hour basis. If the dedicated line is used only for voice or high-speed data transmission, it may be called a "full period" (FP) line, meaning that you have the use of it for the full period of its availability.

If you do not have enough conversation or data to justify the installation of an FP tie line, you may wish to talk with one of the specialized common carriers now offering similar service between certain cities. The rates charged may be less than the interstate rates shown in AT&T Long Lines tariffs. Western Union offers a Hot Line service (see Chapter 14) which connects a special telephone on your desk to a telephone on the desk of your associate in a distant city. You pay for the service by the minute. However, Hot Line cannot be connected to switchboards and dial equipment in Bell System companies, although some independent companies allow such connection. As the FCC is encouraging private carriers to compete with AT&T, you'll find many interesting competitive arrangements will be available to you as time goes on.

## INTRASTATE TIE LINES

These lines are under the jurisdiction of the state utility regulating commission. Rates are usually higher than those applicable to interstate lines. In some states tie-line usage can be arranged for short periods, such as an hour or two daily. If you are interested in "short period" service within a particular state, it is best to consult the tie-line section of your local utility's tariff. This type of tie-line service is not offered in some states.

## TYPES OF TIE LINES

In common with WATS and FX lines, tie lines may be connected from your PABX dial switching system to another PABX dial switching system. Or a tie line may connect a dial system with a manual system, or it may connect two manual systems. Or you may arrange a connection from your system to a specific telephone instrument or from one specific instrument to another specific instrument. Conversation over a tie line flows in both directions, but only one conversation can occur on a single tie line at any one time. If you wish, you can restrict a tie line so that calls can be originated from only one end, not both ends. For instance, a branch store may wish to reach headquarters for credit authorization. A one-way incoming tie line from the branch to headquarters will ensure that the line cannot be used by headquarters personnel to call the branch. When many tie lines are used between two points, some may permit calls to originate from either end; others may be restricted to origination only at one location.

In Chapter 14 I will be discussing the use of tie lines for transmitting written messages or data. You may hear the term "full duplex" when reference is being made to a tie line. This means that the flow of data or administrative messages may occur simultaneously in *both* directions at one time. "Half duplex" means that transmission can be only in one direction at one time. Naturally, when a tie line is being used for voice conversation, each person can talk alternately if the other person is a good listener!

### Two-Way Manual Ringdown Tie Lines

This line runs from one switchboard or console to another one or from a switchboard or console to a key-equipped telephone in a small office. A person wishing to use the line dials the operator and requests it. If it is busy, his name is recorded and he is notified when the line is free. The operator places her cord in the tie-line jack and manually rings the switchboard or telephone at the other end. The operator at the other end follows a similar procedure on a call which originates there. If the line is connected to a button telephone, a ringing button can be provided to the user, or an automatic ringing mechanism can be installed. When the telephone handset is picked up, the instrument or switchboard at the other end is rung automatically.

### Two-Way Dial Repeating Tie Lines

A manual tie line takes operator time. Thus, if many tie lines are required between two points, they are connected into the dial equipment. The user dials a single- or double-digit code to secure the line, and he then dials the three- or four-digit station number which is desired at the other end of the line. The person at the distant end follows a similar procedure to secure the line. In this way, no operator handling is involved. However, if a user wishes to talk with the distant operator, or have her dial a toll call or a local number from the distant end, the user dials the access code, then "0," and secures the distant switchboard operator.

The dial repeating tie line is apt to frustrate an executive if the tie lines are on dial access and there is a heavy demand for the facility. He may make toll calls rather than use the more economical dedicated tie-line facility. Thus, we feel that operator-controlled tie lines serve the user more efficiently unless your firm is willing to install and pay for sufficient dial-access tie lines to minimize busy signals.

### Combination Dial and Manual Tie Lines

Suppose that your Philadelphia branch office has a manual telephone system served by a "number please" switchboard. However, you wish to avoid involving your own New York headquarters operator in handling tie-line calls to Philadelphia. You arrange with telco to put the tie line on one of the dial levels of your PBX system. Perhaps you have no station numbers which begin with "7." Your Philadelphia tie line is connected to that level. When people dial "7" in New York, they automatically ring the Philadelphia branch office operator. She connects the caller manually to the desired person. In Philadelphia, your people call their switchboard operator and ask for the tie line. They now hear your New York PABX dial tone and can dial your three- or four-digit station number without attention by your headquarters operator.

## MULTIPOINT TIE LINES

The design of a tie-line installation or network is a job for experts. A company using a large quantity of tie-line service may be assigned to the local telco's tie-line specialist. In some cases, a representative of AT&T's Long Lines department may work directly with the tie-line user. Tie lines may not radiate like spokes in a wheel from your headquarters location. Instead, they may run in an irregular pattern from city to city. The tie-line experts in your telco or at Long Lines will try to see that the network is adequate for your anticipated traffic. They also will try to arrange the circuits so that the lowest mileage charges apply.

When a single circuit has many different "drops" or points at which conversation can enter or leave the circuit, it is preferable to arrange a dialing code. Thus, your operator does not have to press a button or pull a toggle to give a code "ring" to signal that particular city so that the telephone there will be picked up. Instead, she dials a single- or double-digit code, and the appropriate telephone rings automatically in that specific city. If you are using sufficient tie lines to justify dial access to those lines for your users, they will dial the access code and then the city or directional code.

If your data processing people wish to use your tie lines during the evening hours, you may wish to equip your tie-line cities with special terminals which can be activated to send traffic by your dialing that city's destination code. The data are made ready at the distant point. The operator at the distant point has a special switch which diverts the tie line from operator handling at the switchboard or console to handling at a telephone instrument in the data processing department. The dialing of the code at headquarters activates the terminal to automatically send its data over the tie line. If an entire series of distant points is to be polled for data, special equipment can be arranged to automatically dial the needed codes, pick up the data, and go on to the next location. If staff members at the distant location wish to converse with headquarters personnel before transmitting data, the tie-line circuits can be so used and then transferred to data transmission through the appropriate terminals and interfaces.

The Bell System is not the only supplier of voice and data transmission tie-line facilities. Carriers such as Microwave Communications, Inc., and the Southern Pacific Railroad offer private full-time circuits or shared-time circuits. In some industries such as trucking, regulatory authorities are permitting shared-time usage of tie-line circuits, a practice normally prohibited. Thus, several different private trucking concerns can share a single tie-line network instead of duplicating facilities. As technology progresses, the consumer will find more and more carriers competing to furnish the service desired.

## COMBINING INTERSTATE AND INTRASTATE FACILITIES

As interstate tie-line rates are regulated by the FCC, mileage charges are uniform throughout the country. Rental is calculated on an airline-mile basis. An intrastate tie line is likewise priced on an airline-mile basis, but the rate per mile is usually greater than the interstate rate, perhaps by as much as 20 percent. However, if you can include an *interstate* point in your tie-line network, the lower interstate rate schedule will apply, even if traffic on the line is predominately intrastate in character.

Therefore, knowledgeable companies carefully study their marketing effort and determine if they can find a city in another state where sales calls could be received from customers, dealers, or agents located in that state. The city chosen should be as close to the home state boundary as possible so that mileage charges can be kept low.

The organization selects rental space for its branch office. If the volume of business does not justify a separate office for its exclusive use, desk space may be rented in a secretarial service, which will also furnish telephone answering service. A business telephone line is installed and its number is issued to any authorized callers. The answering service will record the messages. A telephone is also installed which is a terminal on the tie line. The answering service can use this tie line to call or dial the other tie-line points to deliver messages or orders. If 24-hour answering service is not economical, telco can install a switch which will cut the interstate point out of the circuit at times when the interstate office is closed. When the switch is activated, the tie-line traffic flows past the interstate point to the desired intrastate point.

Perhaps an example will clarify this money-saving idea. Let's assume that you have your headquarters in New York City. You have important branch offices in Albany, Syracuse, and Buffalo. You'd normally ask the New York Telephone Company to arrange an intrastate tie-line circuit for your needs. However, you decide that you'd like to give your customers in the Erie, Pennsylvania, area an opportunity to enter their orders by telephone. Erie is an interstate point from New York City. It is also an AT&T rate center, which ensures excellent maintenance of your tie line. Your telco marketing representative will now contact AT&T's Long Lines representative. As Erie is included in the network, your line becomes interstate instead of intrastate. Your rental cost for the line may be 15 to 20 percent less than the cost of a line which connects only intrastate points. The saving will be substantial enough to offset or exceed the space rental and secretarial costs connected with the answering point. However, you should make sure that the location and facilities selected will meet telco's tariff regulations covering the location of an interstate tie-line "drop" or point.

## "CONDITIONING" TIE-LINE CIRCUITS

The words "conditioned circuit" have a data connotation. These circuits are specially treated by the carrier to better carry high-speed data. If you are planning to use your circuit for such a purpose, be sure to secure the required data transmission speed information from your data experts. Conditioning adds to the cost of your circuit. In modern technology, certain data "modems" or interfaces are preconditioned. If such equipment is being used by your people, you may not require specially conditioned lines.

## USING TIE-LINE CIRCUITS FOR REDUCING TOLL COSTS

Should you use a tie line for the first leg of an expensive toll call? The answer is "maybe." Let's suppose that your New York office has a tie line to Chicago. You have two toll calls to make: one to Milwaukee, the other to Los Angeles. You decide you'd like to take a "free ride" to Milwaukee by using your tie line as far as Chicago. If it is a dial tie line, and it permits you to dial "9" when you hear the Chicago dial tone, you can dial the Milwaukee area code and your number. If the tie line is operator-controlled in Chicago, your operator there can dial the Milwaukee number for you. You'll save 55 cents on a three-minute call over the rate you would have paid had you dialed Milwaukee directly from New York. On the other hand, if you use the same procedure for reaching Los Angeles, your saving would be only 10 cents at 1973 rates. You would be occupying the tie line, and someone who may have wanted to call your Chicago office from New York would be frustrated and would dial a $1.15 toll call.

Thus, using tie lines to reduce the cost of toll calling must be examined carefully. Do your economic study, and then instruct your people on the way they can reduce toll costs to specific points through tie-line usage. Be sure to consider the accounting problem on such calls. In the example above, the toll charges for your call will appear on your Chicago telephone bill. Your regional controller at Chicago may insist that time be taken to ticket and time the call, so he can charge your office for this expense.

If you wish to make toll calls on tie lines, you must decide how the mechanics should be handled. If the line is operator-controlled, you must dial "0" for the Chicago operator. That person may dial the call for you on an outgoing switchboard trunk, or she may extend the trunk to you and ask you to do the dialing yourself. If you are using a dial tie line, you secure the Chicago dial tone of your dial switching system. Then you dial "9" for an outside line and the access code, if any, for toll. Remember that any dial-access tie line can be "restricted" so that those using it are denied access to the level which secures an outside trunk. In this way you can restrict the tie line to intracompany use and prevent its being used for making toll calls.

A further problem may arise on clarity of conversation. Each time a call is physically connected through a switchboard jack to another trunk, there is a lessening of sound. Let's assume that you have a tie line from Chicago to your Los Angeles office. Los Angeles has a tie line to San Francisco, and San Francisco has a two-point line to Portland and Seattle. Technically, you can call Los Angeles and ask your operator to connect you to the San Francisco tie line. You ask the San Francisco operator to connect your call to her northbound tie line and give Seattle their special code ringing signal. Can Seattle hear you clearly as you talk from Chicago? Maybe, and maybe not. The Bell System understandably suggests that tie lines be kept free of such traffic. In fact, FCC Tariff 260 stipulates the number of connections which can be made through appropriate interexchange channel switching arrangements. But a cost-conscious company may find ingenious ways of using its expensive tie-line circuitry.

## NEW DEVELOPMENTS IN PRIVATE TIE-LINE PRICING

My readers will realize that a book written about a fast-changing technology cannot quote accurate rates for specific services. At the same time, it is sometimes helpful to give you rate information on both a current and a proposed basis, as such history may help you appraise the properness of rates currently in effect at the time you read this text.

The 1973 AT&T Long Lines rate schedule for half-duplex private line service on an interstate basis is shown in Table 1. Full-duplex service is available for an additional 10 percent. The full-duplex line is used by organizations with a requirement for two-way simultaneous data transmission.

**TABLE 1 AT&T Rates for Half-duplex Lines**

| Mileage | Rate per mile for half-duplex service |
|---|---|
| First 25 | $3.00 |
| Next 75 | 2.10 |
| Next 150 | 1.50 |
| Next 250 | 1.05 |
| Additional miles | 0.75 |

As certain private carriers were offering similar service at lower rates, AT&T proposed a new system of pricing. The new rates were submitted to the FCC in February 1973. Three types of rates were proposed:

1. Where there is "high-density" traffic $0.85 per mile
2. Where there is "low-density" traffic $2.50 per mile
3. Where a short haul of less than 25 miles is involved $3.00 per mile

In addition, the former charge for a single "Service Terminal" charge was split into two sections: a Channel Terminal charge and a Service Terminal charge. If the user decides to route his tie line through a high-density traffic area in order to reduce his mileage cost, he may do so. However, he must pay the terminal costs each time his circuit leaves or joins the high-density or low-density sections of his network. The configuration of the network must be done carefully to ensure the lowest cost.

This development is tentative and must be approved by the FCC before it can take effect. Meantime, you can be sure that the private carriers will also be polishing their competitive marketing techniques. You must be alert for the newest developments!

## MEASURING TIE-LINE TRAFFIC

Once you install one of these expensive circuits, it is important that you get your money's worth. Try to use the circuit for data after your switchboard closes. Arrange with telco to convert an operator-controlled line to dial access when your operator activates a switch at her board. In this way your tie lines can be used after hours.

Periodically, you should ask telco to measure your tie-line traffic in both directions. Telcos use different methods, but equipment is available to count calls and measure talking minutes on any tie lines which are terminated in dial switching equipment. Some telcos may be able to attach pen recorders to the circuits so that you can secure the length of each call.

Once you have this information, you can apply some of the usage standards mentioned in Chapter 3. Are you loading your line with over 8,000 talking minutes monthly? If so, you are getting reasonable value for your circuit. If not, and if your toll statement lists calls to the distant number which you can reach by your tie line, you have an educational opportunity on your hands.

LPLs do not terminate in switching equipment, but instead appear on key-equipped telephones. Perhaps several different brokers have access to the "hot line" which carries orders to the traders. If you wish to measure traffic on such lines, the sensing mechanism mentioned in Chapter 3 may be helpful. You tape the sensor close to the button controlling the LPL. When the button is illuminated, the sensor starts the meter, and you secure the number of calls, total talking minutes, and, optionally, the length of each call. Such a sensing unit can also be used at the switchboard to measure traffic on the tie lines which appear there.

I urge you to regularly measure your tie-line traffic. Tie lines are subject to a minimum usage charge covering only 30 days if they are manual, operator-controlled lines. They can be installed or removed as conditions change. If the tie lines are installed on a dial-access basis, selectors and connectors may be involved in each line. These switches are expensive, and most telcos will want you to sign a contract indicating that you will pay "termination charges" if you remove the line in less than five years. Thus, make a careful traffic analysis before you sign that contract!

## TELPAK CIRCUITS

Consumers of large quantities of tie-line services use Telpak service. Telpak is an offering of the Long Lines department of AT&T. Its technical name is Series 5000. Think of Telpak as a large pipe or conduit through which pass many circuits. In 1973, AT&T offered two Telpak services. Type 5700 gives the user 60 voice circuits in a single package. Type 5800 contains 240 voice circuits. The subscriber may use less circuits than the package contains. The 1973 mileage rates were: Type 5700—$30 per mile per month; Type 5800—$85 per mile per month. Many Telpak users combine their voice and written message needs in a single Telpak conduit. This is because each voice channel may be subdivided into a maximum of two telegraph-grade channels.

Telpak is very flexible, and it may be used as an intermediate service when an organization has high-density traffic between two geographic areas. For instance, a Cleveland company has three different plants in the Detroit area, plus operations in St. Paul, Chicago, and Green Bay. The heavy traffic between Cleveland and Detroit is loaded into a Telpak Type 5700 circuit. It easily handles voice traffic, high-speed data, and some telegraph-grade circuits for order entry. Traffic to the other midwestern plants likewise travels on Telpak as far as Detroit, but then it travels in Series 2000 tie lines from Detroit onward. Series 2001 is AT&T's designation for the normal voice tie-line circuit.

Telpak may be available on an intrastate basis in a package with less circuits than described above. The cost is correspondingly less. Your telco marketing man can give you details.

## BROADBAND CIRCUITS

Your data people may ask you to procure a broadband circuit for data to be transmitted at especially high speeds. Conditioned lines, mentioned earlier, are available in the normal Bell System Series 2000 offering. Series 3000 channels are specially designed for data transmission, and detailed specifications on line conditioning are available from your telco or Long Lines marketing representative. AT&T offers Series 8000 channels with a 48-kilocycle "nominal carrier spectrum" (this is data language to describe bandwidth). Telpak Series 5000 offers 240-kilocycle bandwidth in Type 5700; and approximately 1-megacycle bandwidth in Type 5800. It is important to note that some, but not all, of these broadband services can be used alternately for voice conversation and data.

Western Union and other specialized carriers likewise offer broadband services between certain cities. The Western Union customer may choose between 2- and 4-kilocycle offerings. The service is not "dedicated" on a 7-day, 24-hour basis but is sold on a measured-time basis. You are charged on a per-minute or tenths-of-minute basis. The customer dials the distant connection point in the United States or Canada. International Datel service can be used to connect this service with certain foreign countries. Almost 50 cities had Western Union broadband service available in 1973. If you have only occasional need to speed data to your home office computer, you should investigate such measured-time services. See Chapter 14.

## NEW DATA SERVICES

As this book has been written primarily for those using administrative communications, I have not attempted to include details on the many new developments in data communications, both domestic and international. If this subject is important to you, check with your telco marketing representative and ask about the complimentary seminars sponsored by the Bell System which train people in data communications technology. Also, management education organizations frequently sponsor private seminars in this field. See Chapter 17. Data processing magazines are sources of current information, as are marketing representatives of the communications common carriers. Keep up-to-date! Tie-line services are being continually improved through new technological developments.

## OFF-PREMISE EXTENSIONS

I have defined the meaning of these services at the beginning of this chapter. Here are a few suggestions on the proper use of this facility.

The OPX is very useful in certain cities where message unit charges apply to local calls. A branch office in the same city may originate a large quantity of traffic to and from the headquarters office. If the offices are relatively close together, the mileage charge for the circuit may be quite reasonable. As unlimited calling is available on the OPX, the line can be kept busy all day. No message unit charges apply. Thus, the OPX may be one of your weapons to combat the ever-increasing cost of message units.

On the other hand, if the branch is far distant it may be unwise to install the OPX, as the mileage charge may be much more than the cost of message units which would accrue on the specific volume of calling being done. You must tally calls in both directions for a reasonable period before making a decision on installation of the OPX.

The OPX is used frequently for easy access to the company's dedicated tie lines, WATS, and FX facilities. As the OPX terminates directly in the switching equipment at headquarters, it can easily access the operator so that the dedicated line can be requested. When the special circuits are switched to dial access after hours, the OPX user can dial into the circuits directly. As mentioned in Chapter 3, a WATS line can be connected directly to an OPX on a night connection, giving the OPX station direct access to a specific WATS line for an evening or a weekend.

Marketing men may find the OPX a valuable tool in their effort to stimulate customers to phone their orders. An OPX is installed in the customer's purchasing department. If desirable, the three-digit number dialed by the customer can be connected to a telephone recorder. When used by customers, the OPX should be restricted so that the caller cannot dial "9" for message unit or toll calls.

If you are considering or using Centrex and you have branch offices in your local Centrex area, you may be tempted to pay any applicable mileage charges and install OPX Centrex numbers in your branches. We suggest caution in this use of OPX circuits. Almost all Centrex tariffs require that such branch lines carry the highest Centrex rate in addition to a possible mileage charge. In contrast, particularly if no message units apply in your city, it may be far less expensive to use regular business trunks for the branch office, asking your personnel there to dial headquarters Centrex numbers directly. If more than one line is required at the branch, consider splitting your service between Centrex lines for intercommunication and city trunks for outdialed local calls.

If you are using a number of OPX lines between the building where your switchboard is housed and another location, telco is probably charging you for mileage applicable to *each* line. In certain telco tariffs, a provision is made for an alternative method of charging for the needed cable. The arrangement may be called a "cable-carrying" tariff, or you may find the service under a title like "local channels." In essence, you pay monthly for maintenance of the single cable which runs from your switchboard location to the off-premise location. This cable carries all of your OPX lines. Your local tariff will tell you if your telco offers such an alternative to OPX individual line mileage charges.

A second alternative applies if the two locations are very close together—separated by an alley or city street, for instance. If you will examine telco's mileage tariff, you may find that a lower mileage rate applies to buildings located on the "same property" than for carrying a line to "different property," or "noncontiguous property," as it is sometimes described in the tariff. To convert the subsidiary location to "same property," employ a contractor to construct a conduit or tunnel between both buildings. As your property is now contiguous, it is possible that there will be no mileage charge at all or that the lower "same property" rate will apply.

Here is a third suggestion. Count the number of PABX stations located in the building which houses your switchboard and switching equipment. Then count the number of OPX stations at the subsidiary location. If there are more OPX lines than main station lines, consider moving your switchboard and switching equipment to the subsidiary location. Your former OPX lines would then become main station lines, so no mileage charge would apply. You would have fewer lines running as OPX lines from the subsidiary location to the main location, and on them you would pay a mileage charge.

## LOCAL PRIVATE LINES

The LPL is sometimes installed for prestige reasons. It may save a few seconds of dialing time, because the called telephone rings as soon as the user picks up his receiver. Thus, the executive may feel that the cost of the line is inconsequential. The analyst should make sure that the LPL cost is justified in comparison to dialing the call through the PABX system.

Most LPLs travel short distances. For instance, a local brokerage office may be generating sufficient business in the municipal bond department to feel that a "hot line" should be established with a bond specialist located in another local brokerage office. Instead of installing a tie line on the switchboard or in the dial equipment, the firm installs a desk-to-desk "hot line" which can be used only by the two individuals involved or perhaps by a number of individuals each with an appropriate button on their multibutton telephone instrument. Such a line can't work as hard as a line which is available to all personnel, because it may be idle when the market is closed. We suggest that such a line be terminated not only on the desk of the person involved but also on the switchboard. Then, when the municipal bond department volume slacks off after trading hours, the line can be made available to others.

LPLs are quite common between investment bankers and local bank trust investment departments. The banker can pick up the hot line and immediately deliver his buy or sell order. Likewise, investment bankers may install these lines between their offices and the offices of other investment banking firms to expedite the transaction of business when the market is active. If this is done, the cost is usually shared between the two firms.

On balance, make sure that the LPL is genuinely useful and that its cost is justified by its utility. As LPLs travel short distances, mileage charges may not be great. Also, if the LPL is used in a city where message unit charges apply, the LPL will save the cost of the message units which would have applied if the call had been dialed through the local outdial trunks (see Chapter 2).

## CHOOSING THE DEDICATED FACILITY WHICH IS BEST FOR YOU

In Chapters 3 and 4 and in this chapter you have read about the special lines which you can install to reduce your long distance costs and message unit charges or to provide special services for you. The following summary and illustrations may help you visualize each service.

### WATS

WATS is a one-way service, inward or outward, which gives you access to many telephones in an area comprising several states (Figure 1). In general, it costs more per line than the other dedicated facilities, but it gives you far greater coverage.

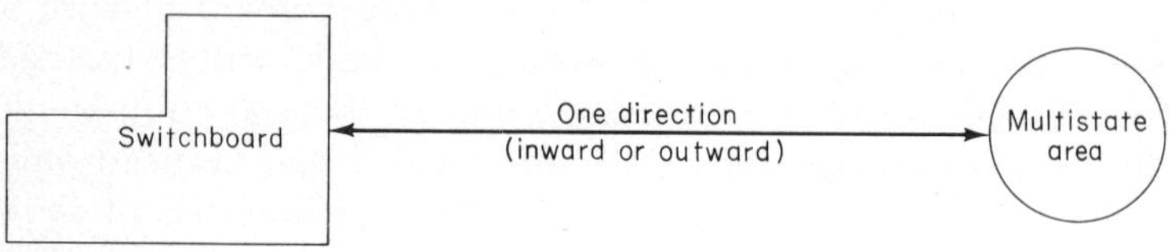


**Figure 1**

### Foreign Exchange

A telephone number from a distant point is brought to your switching equipment or telephone (Figure 2). When you use your FX trunk, the dial tone you hear is that of the distant city. You can dial any number in that area, and you may be able to access long distance circuits if you desire that service.

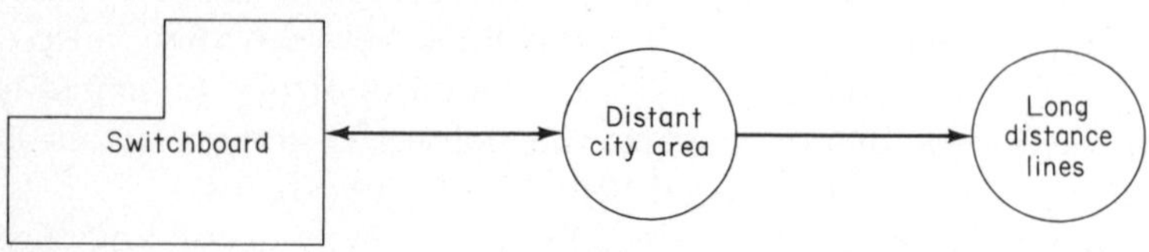


**Figure 2**

## Tie Line

This service connects your switching equipment or telephone with another specific location in the distant city (Figure 3). If you wish to call another number than that of the specific location, you must either dial "9" when you hear the distant dial tone or ask the distant switchboard operator to do the dialing for you.

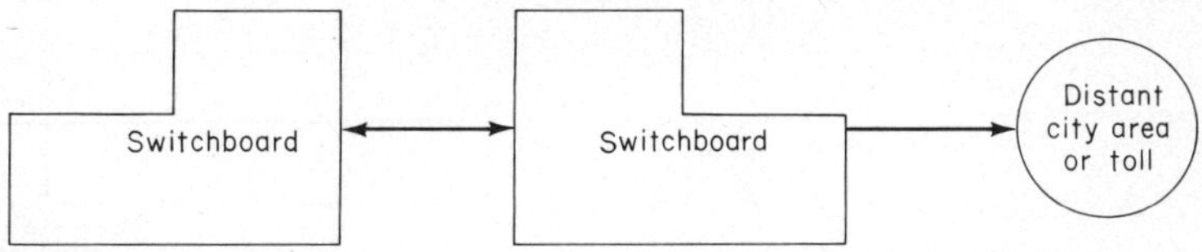


**Figure 3**

## Off-Premise Extension

This line runs from your dial system to another location (Figure 4). Your people can dial that location directly without utilizing a central office trunk. Similarly, those at the separate location can dial any station number at the main location, and they can dial "9" for an outside trunk or dial other access codes which permit their use of tie-line, FX, or WATS facilities.

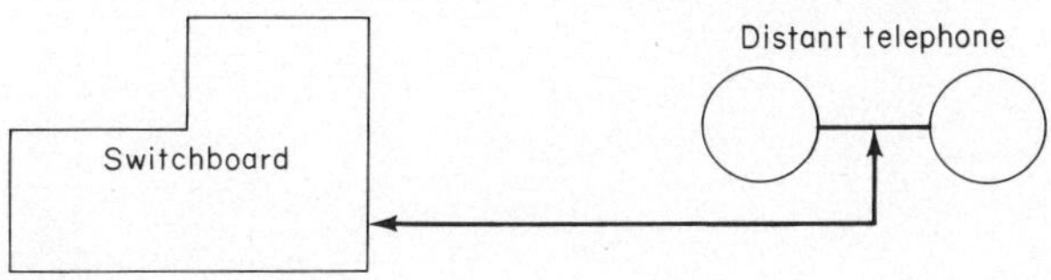


**Figure 4**

## Local Private Line

This line runs from desk to desk in your building or to some other location within the area served by your local telco (Figure 5). If desired, the LPL can appear on your switchboard so that the operator may control its use. It may be arranged as a "hot line" so that the distant telephone rings as soon as you lift your telephone receiver.

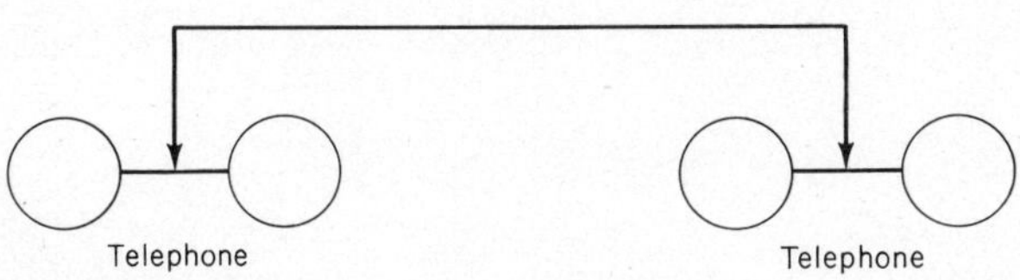


**Figure 5**

## Cost Calculations

Your telco marketing representative will itemize the cost of each alternative. WATS costs are in the public tariff. The cost of other dedicated facilities is based on the airline mileage between the points involved, plus monthly rental charges for the connecting and switching devices associated with the circuits. The 1973 interstate mileage rates are given in Table 1.

CHAPTER SIX

# SUGGESTIONS FOR READING THE UTILITY'S TARIFF

In the following chapters, I will be describing equipment which is supplied by the utilities to users of telephone and telegraph service. As all utilities are "regulated" industries, the services they supply and the prices charged must be a matter of public record. The document in which the services are described and priced is called a "tariff." The tariff is a legal document which could be considered the consumer's guide to telephone and telegraph service. Thus, if you are using a substantial quantity of the utility's service, it is *very* important for you to become familiar with the public tariff.

For instance, the tariff will tell you if certain equipment, such as a telephone system, switch, console, or switchboard, is subject to a binding contract which requires rental payments over a defined number of years. If you should require a different system before the end of the contract period, you may be required to pay all or a portion of rental charges covered by the contract even if you select a new system with higher rental charges. The tariff will tell you how much it will cost to install equipment, to change it, or to move it. If you must run a telephone line to a branch office location in a suburb of your city, the tariff will tell you the cost per mile of that off-premise line.

Most important, the tariff will list systems, instruments, or circuits that will be alternatives to those which may have been recommended by the marketing representative of your utility. You may find a system which will give you the needed features at a decreased rental cost. Or the tariff may describe a new pricing system in which you pay more rental in early years and less in later years, or vice versa.

Telephone and telegraph tariffs are always found at the principal business office of the utility. The thick volumes are usually stored in a location which is convenient to customers who visit the business office to arrange for service or to pay bills. As the tariff is a *public* tariff and must be shown to any member of the public on request, you will have no difficulty in procuring a copy for examination. Just ask a member of the business office staff to show you the tariff. Usually it is produced within a few minutes, and you are ushered to a convenient desk or table where you can study it. It is *not* necessary to ask for permission to examine the tariff, nor is there any requirement that a member of the utility's staff stand by to supervise your tariff reading.

Let me give you a few illustrations of situations where tariff inspection was of great help to the companies involved.

## CHOOSING A NEW TELEPHONE SYSTEM

A firm in Wisconsin served by an independent telco was contemplating replacing an ancient manual system. The deluxe package suggested by telco had many fine features. But a visit to the local business office showed that the tariff indicated two other systems were available—with savings of 30 and 45 percent, respectively, over the quotation already submitted. The customer was delighted to discover this and did *not* accept the deluxe package.

## ADDING ANOTHER SWITCHBOARD

A company in the South had reached the capacity of its two-position cord switchboard and needed another position. Telco proposed to replace the old switchboard with a modern 608-type board. The company was impressed with the features of automatic recall and ringing, low profile, digital clock, beige color, and other advertised advantages of the "Universal" switchboard. The price tag was in the $150 range for each of the three required positions. Telco explained that the older-type switchboard was no longer manufactured by Western Electric and urged acceptance of their proposal for the newer model.

A visit to examine the tariff seemed in order. In the fine print of the section on PBX systems it was stated that the older-type switchboard was indeed "manufacture-discontinued." But the fine print went on to say that the board could still be furnished to *expand existing installations*. The cost? $50 monthly. And an optional package was available which would give automatic ringing and automatic recall for another $35 monthly. The customer would have the basic advantages of the 608 board for a total monthly rental cost of $85 versus $150. What did he do? He was worrying about increasing the earnings per share on his common stock and wasn't about to pay almost double the cost for decor and some digital clocks.

## REDUCING MESSAGE UNIT EXPENSE

A firm in a Philadelphia suburb was concerned about excessive message unit expense on suburban trunks. An examination of the relevant tariff showed that it offered both flat-rate trunks with unlimited local calling privileges at a higher monthly rental and message-rate trunks at a nominal fee but with message units applicable to all calls, including multiple message units on calls beyond the limits of the suburban exchange. Thanks to a conscientious telco marketing representative, it was discovered that Pennsylvania Bell permitted the *mixing* of the two types of trunks in one switchboard installation. What did the customer do? He rented message-rate trunks at low cost for the incoming lines: obviously, there is no message unit cost on an incoming call! And he rented flat-rate trunks for the use of his outdial callers. The savings? Substantial!

## KEEP CLOSE CONTACT WITH TELCO

Really, it *pays* to read the tariff, and, believe me, it also pays to stay on the right side of your telco marketing people. They really want to help you if you'll tell them what you want and if you will educate yourself enough to tell them *clearly* and in no uncertain terms.

Am I convincing you? The public tariff is a treasure trove of information for the serious-minded telecommunications administrator. It's readily available. Sometime, when you have lunch with your telco marketing representative, ask him if he can duplicate copies of the particular tariff sections which interest you. However, telco is *not* obligated by law to furnish you with copies. If you want photocopies, your telco may duplicate the pages you want at a fee. In every state capital there is a service which follows tariff changes filed by the utilities with the state public utility commission. You can subscribe to such a service and receive copies of each tariff filed. You may find that your city's public library keeps a copy of telco's tariff. If so, copying equipment is readily available there. Otherwise, take plenty of paper and your pen and copy the information you need as you examine the tariff.

## SOME SUGGESTIONS ON EXAMINING THE TARIFF PAGE

The tariff page reproduced here as Figure 1 is quite typical of those you will find in your own telco's business office. Please note the following features—and when you are copying from the tariff, be sure to record the date the tariff became effective, plus the section and page numbers, so that you have concrete evidence to back up the orders you will be giving your telco representative. Here are common elements of most tariffs and tariff pages:

1. *Index.* Look in the first volume of the tariff for the index. Are you interested in key systems? PBX systems? Mileage? Special equipment? Installation, move, and change charges? Call Directors? The index will usually direct you to the proper section number.

2. *Section number.* The tariff is always in numerical sections, but frequently you can speed finding your section by looking for the index tabs that may separate each section. The section number is printed on each page of the tariff in the upper right corner.

3. *Sheet number.* Tariff pages are usually known as "sheets." Also, as changes are made, a "Revised Sheet" is inserted. If you find such a sheet with a reasonably current date on it, you may be sure that something in that tariff page has been changed recently. Usually the next line in the upper right corner says something like "Cancels 2nd Revised Sheet" (see Figure 1).

4. *Issue date.* Look carefully: the sheets are issued well ahead of the effective date, so you may have a chance to save money if something is about to increase in price!

5. *Effective date.* This important date may appear at the top of the sheet, as in Figure 1, or at the very bottom, usually in the right corner.

6. *Paragraph numbering.* Be very careful. Read before you write! Services are listed under commercial, residential, Centrex on your premises, Centrex on telco's premises, different rates for different

Northwestern Bell Telephone Company
State of Minnesota
Issued: May 18, 1972

GENERAL EXCHANGE TARIFF

Section 33
3rd Revised Sheet 8
Cancels 2nd Revised Sheet
Effective: June 1972 Billing Dates

CENTREX SERVICE

C. RATES (Cont'd.)

2. Stations (Cont'd.)

a. Commercial (Cont'd.)

(2) CENTREX II

| | Monthly Rate | | | |
|---|---|---|---|---|
| | First 100 Stations | Next 100 Stations | Next 700 Stations | Over 900 Stations |
| (a) CENTREX stations, Per station* | | | | |
| Switching equipment located | | | | |
| In customer premise | | | | |
| Rate Groups I-VIII | | | | |
| Exchange Access | $ 4.55 | $3.60 | $2.65 | $2.10 |
| Intercommunication | 11.80 | 7.55 | 6.05 | 5.05 |
| Rate Groups IX-XIII | | | | |
| Exchange Access | 6.00 | 4.70 | 3.45 | 2.75 |
| Intercommunication | 11.90 | 6.45 | 5.50 | 4.60 |
| In Tel. Co. premise | | | | |
| Rate Groups I-XIII | | | | |
| Exchange Access | Same as above | | | |
| Intercommunication | Same as above, plus $.25 per station | | | |

(T)

(T)

| | Monthly Rate |
|---|---|
| (b) Restricted stations, each | same as "Over 900 Stations" rate |
| (c) Extension stations, each | |
| CENTREX station extensions | $5.90# |
| Restricted station extensions | 5.90# |

\# Applies to extensions located in same building as associated CENTREX or Restricted station. Extensions may be located in a building other than that of the associated CENTREX or Restricted station subject to mileage charges as specified in Section 10.

(d) Service connection, move and change charges are as provided in Section 12 of the Tariff

*Total rate per station equals Exchange access plus Intercommunication increment (T)
(T)



**Figure 1.** A sample tariff page. *(Reprinted with permission from Administrative Management)*

parts of your city, and similar complexities. On the sample page reproduced here, notice that we are in:

Section 33: Centrex Service
Subsection C: Centrex Rates
Sub-subsection 2: Centrex Stations
Sub-sub-subsection a: Commercial Centrex Rates
Sub-sub-sub-subsection (2): Centrex II Service
Sub-sub-sub-sub-subsection (a): Centrex II Station Rates

7. *Special symbols.* In the page illustrated, we have only one asterisk to interpret, and its explanation is reasonably clear. However, the normal page may have five or six special symbols which

are footnoted. Always be careful to read the footnote with the original text. Otherwise you may read, for example, "manufacture-discontinued" without noticing the footnote which indicates that reconditioned units from "Class C" warehouse stock are still available to enlarge present equipment.

8. *Marginal letters.* In Figure 1, notice the (T) to the right of the Centrex pricing information. This means that the text wording has been changed since this particular sheet was last issued. The footnote likewise has a (T). It is evident that the Internal Revenue Service (IRS) tax change on Centrex was responsible for this change in Northwestern Bell's mechanical setup of Centrex charges.

## APPLYING IDEAS GAINED FROM TARIFF READING

Now that you have studied the typical tariff page (Figure 1), let me explain how you can use the information you have received.

Notice the way Northwestern Bell has broken down its Centrex tariff into two sections. You'll note that one section is called "Exchange Access" and the other one is "Intercommunication." This is a new way of expressing the cost of Centrex lines. An interesting story lies behind it, and most Centrex users have now received substantial refunds from their telcos.

As Centrex users soon discover, the federal excise tax applies to all numbered Centrex lines. Formerly, any new Centrex user suddenly found himself paying a greater federal excise tax on his Centrex system than he had paid on the traditional dial system which was replaced by Centrex. Usually, the extra tax cost had not been included in the economic study which preceded the adoption of the Centrex proposal. Many years ago, the IRS decided that the consumer should get a break. The IRS indicated that if telco wished to do so it could divide the charge for a primary Centrex line into two portions. One portion could be considered straight intercommunication within the office itself. The other portion could be considered the part of the line, timewise, which is used for communication with the outside world; only this portion would be subject to the federal excise tax. In the page illustrated, the amount of tax applicable to the primary Centrex line was reduced 66 percent from the former figure.

So, you can see that had you read this particular tariff page you might have been curious about the new way that telco expressed Centrex pricing. Your asking telco about this tariff change could have brought about a speedup of any refund due you for overpaid taxes which accrued between the time that your particular telco decided to act on the IRS decision and the time that the accounting department actually adjusted your tax bill.

Please glance once more at the sample tariff page. In the center of the page after heading (b) you'll note "Restricted stations" and a reference which indicates a cost of $7.35 monthly for such a Centrex line in rate groups IX–XIII. You've just discovered another reason for tariff reading: the existence of a class of Centrex service which may not have been explained to you when you considered or installed your Centrex service. "Restricted" Centrex is an intercom line and instrument which has its own four-digit number. It can call any number within the Centrex system or be called by any other user within the system. But it cannot make outside calls or receive such calls. Every Centrex user can find employees whose job responsibilities require calling only within the Centrex system. In Minnesota, the fully restricted line saves 29 percent of the monthly cost of the least expensive unrestricted line. That's a cost reduction which would appeal to any controller. If you are a Centrex user and you haven't heard about this option, may I suggest you call your local telco marketing man and place an order?

## FINDING OUT ABOUT NEWLY FILED TARIFFS

Because all telephone service provided by the utilities is regulated by some governmental body, a new tariff must be submitted to that body for approval. Further, most state laws require that public notice be given at the time a new tariff is filed.

It sometimes takes a very good detective to discover where the public notice is posted. In one New England state, the notice is sent to the town clerk, who is supposed to post it on the town public bulletin board. In other states, publication in the daily legal newspaper is sufficient. Occasionally, enterprising reporters will discover that the rates are about to be increased and publicize the fact in newspapers of general circulation. You can easily find out how *your* state's new tariffs are publicized by asking the secretary of your state public utility commission to inform you. You can then subscribe to any needed publication or perhaps examine it in your public library.

In some states, such as Ohio, the telco may file the full text of the new tariff in the public business office at the time the tariff is presented to the public utility commission. If this is true in your state and if a rate increase is pending, ask the person at telco's business office counter if you can examine the proposed tariff as well as the existing tariff. That way, there will be less guesswork as you study the economics of buying or renting an interconnected telephone system to protect you against the utility's future rate increases. Those increases *must* come about so that your utility can cover its increased costs for labor and materials.

Is tariff reading worthwhile? You can be sure it *is*. Let's put it this way: your telco has problems common to any business. It has a constant turnover of personnel. The man who takes the place of your recently promoted marketing representative may be fresh out of training school. Once you see the thick volume in which tariffs are contained, you can understand why even an experienced telco salesman won't know all the details about a particular piece of equipment or service which might be made to order for your particular problem.

I'd like to emphasize again how important tariff reading is. Find out where your telco's tariff is housed. Reserve a morning or afternoon for some leisurely reading. Then surprise your telco representative with your new knowledge about the complexities of his profession.

CHAPTER SEVEN

# TELEPHONE SYSTEMS FOR THE SMALL BUSINESS

Has your sales vice president just announced that your company will be opening a branch in Atlanta? Does the Los Angeles district manager complain about busy signals experienced by customers when they dial his number? Perhaps you are the chief executive of a small business which needs only a few telephone lines. You are wondering if you should be investigating a small dial system which would permit you to answer incoming calls and at the same time provide intercommunication with others in your office.

These problems involve the selection of the right telephone system for the small business. Some systems comprise nothing more than a six-button telephone on each desk. Others require larger multibutton instruments. Another office may be a candidate for a manual switchboard or a small package dial system rented from the utility or purchased from an interconnect vendor.

Here are some suggestions on furnishing the small office with equipment which will provide a proper grade of telephone service. Additional information on key systems and multibutton equipment is found in Chapter 11.

## MANUAL SWITCHBOARDS

Years ago, the words "Number, please?" were heard throughout the land. Even today, in some rural areas dial telephone service is still a development of the future. But many businessmen do not realize that the manual switchboard is still very much alive. It is inexpensive. It has adequate capacity for any reasonable number of central office trunks and stations to serve the office workers.

The best way to find out about the manual switchboards offered in your locality is to examine the tariff of your utility in the business office (see Chapter 6). Western Electric has introduced a console-type manual switchboard in recent years, but it isn't offered by every Bell System company. Some Bell System companies have developed their own switchboard instead of using the Western Electric product. The independent utilities offer equipment which is very similar to the Bell System units.

If an actual switchboard is required because your office needs exceed the capacity of multibutton instruments, the most common unit is a shelf-type manual switchboard with the usual cords, jacks and toggles, or switches for ringing the telephones in the office. In most telcos, rental charges are very reasonable, perhaps $25 monthly. The unit is installed at right angles to a clerical desk. The clerk serves as a switchboard operator when needed. When an incoming call is received, an audible signal is heard. The operator inserts the cord in the trunk jack and connects the other cord to the station jack of the individual wanted. She then rings the telephone manually. The audible signal and supervisory lamps on the switchboard will tell her when the conversation is finished. If a call is to be transferred, the person called will slowly depress the carriage button on his telephone. The audible signal and illumination will tell the operator that her attention is needed.

Intercommunication among the members of the office staff is possible by picking up the handset and waiting for the operator to respond. She then connects you to the person you want. The small dial intercom systems described in Chapter 11 can likewise be used within the office for dial intercommunication service. Such systems are faster than waiting for the operator to make the connection.

When a member of the office staff wishes to make an outside call, he picks up his handset and asks the operator for an outside line. To expedite such service, we suggest that he dial "0" on his telephone instrument. This puts a flutter on his switchboard lamp as the dial revolves. The flutter tells the operator: "Give this man an outside line." She connects into one of her central office trunks without need for conversation.

### Private Business Lines

To expedite outside calling, some manual switchboard users install private business lines at the desks of people who make many outside calls. If desired, those numbers can be billed separately by telco, and each private line user receives his own telephone bill which itemizes message unit and toll expense. Such private lines may also give the user direct inward dialing service, bypassing the switchboard. This service may be useful if a manager wants to encourage his traveling staff to call him on a station-to-station basis at the lowest cost. If the private line is not answered, it is evident that the executive is not there and the traveling employee does not have to pay for the call. If the line is busy, the traveler knows that he should wait and try again in a few minutes.

Here is one caution on the use of private business lines in any telephone system, manual or dial. If these lines are installed for one executive, others may wish this convenience, too. In areas where unlimited local calling is allowed, private business lines may be very expensive. They serve a business much more efficiently when they are installed on a switchboard. If installed at an individual's desk, they should at least be shared with other individuals to maximize usage.

Another problem involving private lines concerns the policy of some telcos to include a message unit allowance in the cost of a business private line. For instance, a $10 private business line might include an allowance of 80 message units. If each line is billed separately, users of that particular line must make sufficient local calls to use up the allowance in any one month. In contrast, if all the private lines are "group-billed" under the main telephone number, their message unit allowances are merged and are available to reduce the company's total monthly message unit billing. But if the lines are group-billed, no separate toll statements will be issued and toll calls will appear on the regular company toll statement without any reference to the private line number. These problems should be carefully considered before the private line arrangements are made.

## MULTIBUTTON INSTRUMENTS

In Chapter 11 I describe how these instruments are used by other persons for answering telephones when the called person is away from his desk or talking on another line. The multibutton telephone, sometimes called a key-equipped telephone, can be an excellent instrument for furnishing service to the small office. The description which follows is based on instruments available generally throughout the Bell System; their equipment is shown in Figure 1 of Chapter 11. Quite comparable instruments are available from the independent telephone utilities. Interconnected systems will be described later in this chapter.

### The Six-Button Telephone

This instrument provides a hold button and five line buttons. Thus, it can answer up to five trunks from the central office. Each individual in the office who requires a telephone is given a six-button instrument. If the office is small, the answering clerk can look around and select some person who can handle the incoming call. She says: "Mr. Miller, please pick up the call on line 2."

If the office is too large for such verbal notification, a small voice intercom can be purchased from one of the electronic supply houses. The master station is located at the clerk's desk. She presses a lever and pages the appropriate person through the speaker mounted on each person's desk. More expensive intercoms may announce the call to a specific desk without the broadcasting operation.

If your need for incoming lines does not exceed four central office trunks, plus a hold button, the sixth button on your telephones may be used for a dial intercom (see Chapter 11). The operator puts the incoming call on "hold." She presses the dial intercom button and dials a single digit. Up to nine persons can be reached by typical systems supplied by Western Electric. Other telephones can be added to the dial intercom system if the additional people are willing to share their dial intercom number with someone else. The operator announces the call and instructs the person which line button to depress for pickup of the incoming call. A dial intercom announces a call by sounding a buzzer mounted inside the telephone instrument. Thus, the dial intercom has the capability of overriding an ongoing conversation. The user hears the buzzer and places his existing call on hold or terminates it so he can listen to the instructions from his answering clerk-operator.

### Two Six-Button Telephones

In almost every situation, a telco marketing representative will suggest a single multibutton instrument of greater capacity if you need more than the four or five incoming trunks just discussed. A second six-button instrument, mounted side by side with the first instrument, may be a less expensive alternative. Again, it will take a study of your local tariff to determine if you can make a substantial cost reduction by such instrumentation. If the arrangement saves money, you will now have a facility which will accept eight or nine trunks and one or two dial intercoms.

As your needs increase, telco may suggest a two-digit intercom so that more individuals can be reached by your clerk-operator. An intercom providing two talking paths and two-digit dialing is even more expensive. Please note that two separate single-digit nine-station intercoms, one installed on each of the two six-button telephones, will give you the ability to reach 18 different individuals. If you have more than that number of telephone instruments in your office, you may shortly be needing one of the small package dial systems which are available from the telephone utilities and the interconnect suppliers.

### The 10-Button Telephone

This instrument is a comparatively new product of Western Electric. It may not be offered by all Bell System companies. It has a capacity of nine trunks or any other configuration suitable to your needs. Its cost should be compared to the cost of two six-button telephones and to the cost of the alternatives next described.

### The 12-Button Key Strip

This instrument is a rectangular box with 12 buttons mounted in it. Excluding the "hold" button, it gives you an 11-line capacity. It may not be offered by some Bell System companies which are offering the new 10-button telephone. It is sometimes known in telephone language as a "KU-2," an "S-6051 key set," or a "Key Chief." The instrument uses a separate telephone instrument mounted on the user's desk. The unit itself may be mounted on the edge of a desk, in the kneehole, or on a triangular base for desk-top use.

### The 18-Button Call Director

This Western Electric product is used most often for the small business which requires switchboard service but does not need a small dial package system. It gives a capacity of 17 lines plus hold (see Figure 3 of Chapter 11). In most tariffs it is more expensive than the alternatives we have described. In some telcos, the standard 18-button Call Director may be modified to a 12-button size and, in certain companies, the instrument is used for 6-button service.

### The 20-Button Telephone

This new Western Electric product, illustrated in Figure 2 of Chapter 11, is exactly the same as the 10-button instrument except for its larger capacity and rental rate. It may not be offered by your telco.

### Other Call Directors

In most Bell System companies, a 30-button Call Director is available, plus a boxlike instrument which contains 60 buttons. In some companies, Call Directors are priced in modules of 6-button strips. Your local tariff will give you details.

## DISCUSSION OF YOUR ALTERNATIVES

In all the telephone utilities, Bell System and independent, there has been a dramatic change in the installation, move, and change charges for telephone instruments. In some companies, these charges have tripled in the last few years. Thus, in considering your alternatives, be sure to check your local tariff and determine what it will cost to install, move, or change instruments once they have been selected. In general, 6-button instruments and the newer 10- and 20-button instruments carry lower installation charges than the 12-button key strip and the Call Directors.

In considering the use of the second six-button telephone, be sure to check your local tariff to determine how you will be charged for the pickup buttons and light bulbs which will appear on the two six-button telephones. For instance, in one telephone tariff, each pickup button and each neon lamp is charged at 37 cents. Then, there is a rental charge of $2.09 for the bridged telephone, assuming that the first six-button instrument is furnished to the answering clerk in any event and that its cost would be present regardless of the clerk's responsibilities for monitoring the lines of other persons. In this same company's tariff, the Key Chief instrument is billed to the user at $9 monthly. If this alternative is chosen, there is no charge for the pickup buttons and neon lamps. Thus, you would have to calculate which alternative would bring you the lowest cost. If you use two six-button instruments to monitor 10 lines, you will pay $2.09 for the additional bridged telephone. There will be two hold buttons, whose cost is 37 cents each. You will have 10 pickup buttons and 10 neon lamps. Your costs are:

| | |
|---|---|
| Bridged telephone | $ 2.09 |
| Ten pickup buttons | 3.70 |
| Ten neon lamps | 3.70 |
| Two hold buttons | 0.74 |
| Total | $10.23 |

You can secure the Key Chief for $9 monthly and have the convenience of the single instrument. On the other hand, if you only need eight lines of monitoring capacity, the economics may change. In the example above, you'd need two hold buttons, eight pickup buttons, and eight neon lamps plus the bridged telephone. Your cost would be $8.75.

Also, keep in mind that you must provide a multibutton instrument to all those requiring telephone service in your office. The resulting total expense may be far more than the cost of using the "Number, please?" manual switchboard described previously. True, the switchboard requires that your clerk-operator position be manned nine hours a day. However, the cost reduction may justify your ignoring the convenience of universal answering of incoming calls at any desk. This is one of the chief advantages of using the multibutton instrument in the small office.

## INTERCONNECTED TELEPHONE SYSTEMS FOR THE SMALL OFFICE

In a recent advertisement a manufacturer of interconnected telephone equipment said that businesses with less than 50 telephone instruments make up 95 percent of United States industry. Naturally, the independent manufacturers see a marketing opportunity to improve telephone service by providing equipment with features heretofore unavailable from the telephone utilities. In a very short time, dependable interconnected equipment for the small office will be available in almost every major city. The equipment will be well designed, easily repaired, and expandable through module construction. One such instrument is illustrated in Figure 1.

One typical manufacturer offers such systems with capacities of 5 to 20 central office trunks. Intercom systems of up to 30 stations are available as built-in parts of the system. Two talking paths are available on the intercom if desired. Systems for 5 or 10 central office trunks can be expanded to the 20-trunk capacity by adding modules.

In addition to the standard features common to the systems we have described earlier, interconnected systems may include features like these:

**Figure 1.** A key telephone instrument for a small business furnished by an interconnect manufacturer. *(TIE/Communications, Inc.)*

1. Conferencing: telephones can be tied together for conference calls.
2. Busy line field: the operator can tell at a glance if a particular station line is busy.
3. Direct station selection: the operator can connect calls immediately by pushing a button.
4. Paging: the instruments contain speakers for paging, or external speakers can be mounted within the office.
5. Camp-on: the operator can notify a user of a waiting call if he is talking.
6. Automatic privacy: no other instrument can listen to an ongoing conversation on a specific instrument unless the user wishes another person to listen.
7. Restriction of outgoing calls: this can be arranged by line or by instrument.

When proper servicing arrangements are available in a specific city, the interconnected systems may provide a very interesting cost alternative over the traditional system. Further, you may be sure that the telephone utilities are awake to this competitive challenge and will be competing vigorously to bring you similar systems.

Here is a word of caution on the acquisition of any kind of interconnected equipment. Such systems are sold outright or leased to you on a basis which returns the full purchase price to the lessor over the period of the lease. Smaller offices usually occupy space in buildings which may not belong to the parent company. Thus, before you acquire an interconnected system, check your lease and ask your office manager if moving plans are being made to take effect at the end of the lease. All interconnected systems are portable and can be reinstalled at the new location. However, cabling of the system may be quite expensive. If the lease is about to expire, it may be better to rent a system from the telephone utility for the remaining lease period and consider the interconnected system when your moving plans are complete.

# CHAPTER EIGHT

# DIAL TELEPHONE SYSTEMS FURNISHED BY THE UTILITIES

Your business has been growing. Your single switchboard operator has been with you for years. She manages to keep up with your increased traffic, but you recognize that you will shortly need another operator and another switchboard if you are to give prompt service to your callers. What should you do? Are there newer switchboards available which will speed the connection of your calls? Should you change your telephone system completely? Should you ask an interconnect vendor to send their salesman to discuss your problem?

The phrase "dial telephone system" may be somewhat misleading when you recognize that many organizations are using telephones equipped with buttons instead of dials. Also, there are still a few large organizations such as hotels and hospitals which are served by the "number, please" manual telephone systems described in Chapter 7. The *tariff* will tell you what telephone systems are offered in your locality by the utilities (see Chapter 6). Your interconnect vendor will give you information on the features of private systems (see Chapter 12).

In analyzing your needs and contrasting your alternatives, it is important for you to understand the several different methods of pricing telephone equipment. Your particular utility may offer only one method, or perhaps its tariff will give you a choice of pricing method.

## UNIT PRICING OF TELEPHONE SYSTEMS

A dial telephone system contains several major components. First, power must be supplied. The mechanism involved is usually called "common equipment." The dial system is operated by a package of equipment. The equipment may be a step-by-step system, a crossbar system, or an electronic switching system (ESS). Next, stations are needed for each telephone user; they are sometimes called "extensions." A bridged telephone is also sometimes called an "extension," as the bridged telephone extends another person's line to a second telephone instrument so that two or more people can answer an incoming call on a single numbered line. Finally, switchboards or consoles must be furnished to answer calls.

Under unit pricing each of these components carries a specific monthly rental cost. In the older step-by-step systems, each separate switch within the equipment may carry a stated rental rate. Switches include line finders, selectors, and connectors. Every telco will furnish an itemization of equipment which lies behind the lump-sum billing for "local service" on your monthly bill. I think it is essential that you secure that listing regularly, as it will tell you exactly what equipment you are renting and how the equipment is priced.

Under unit pricing the *grade of service* is under the control of the customer. Suppose that you receive complaints that your people cannot secure a dial tone when they lift their telephone receiver. This is an indication that you may need additional switches. Your utility will ask their traffic engineers to review the situation. If you are a Bell System customer and you do indeed need additional switches, they will be ordered from Western Electric, which will also install them. As telco has to purchase the switches to be added to your system, they may ask you to sign a contract. The contract provides that you must pay a defined amount over the contractual period—usually five years. This ensures that telco recovers its investment, even if your traffic decreases and the switches are no longer needed. I will be giving you some suggestions on avoiding this type of problem in Chapter 11. However, the important point about unit pricing is the fact that you, as a consumer of telephone service, can decide what grade of service you wish to extend to those using your dial telephone system.

## SERIES PRICING OF TELEPHONE SYSTEMS

The series pricing concept changes the method of billing you for the telephone system you select or are using currently. The Bell System has encouraged its sister companies to change their pricing systems to the series pricing concept. In this newer system, the consumer tells his telco marketing representative the quantity of numbered station lines he requires to serve his people plus the number of bridged telephones for individuals who have light telephone traffic. The consumer likewise indicates the growth he expects in the system within the next five years. If the consumer wishes certain special features included, he so instructs telco's salesman. A recommendation is then made by telco. Usually the recommendation suggests a single system which presumably meets the consumer's requirements. In this pricing method, charges for power and switching are included in the rate for each telephone station. Supplementary charges are made for bridged telephones and the needed switchboards or consoles. Key equipment is also billed separately.

Thus, responsibility for providing the correct number of talking paths, which in turn are provided by switches, is passed along to telco. If troubles develop, it is telco's responsibility to correct the problem *at no charge to the consumer.* If your telco is operating on the series pricing concept and you are having problems, be sure to insist that your problems are corrected promptly.

If you are in the market for a new telephone system, I urge you to read your telco's tariff. Otherwise, you'll be unable to appraise any telco proposal which recommends a single telephone system. It would be better if you would instruct your marketing representative to bring you perhaps three alternatives so you can examine the different cost and service features.

Another caution: be sure to estimate your growth potential generously. Telephone systems supplied by all utilities and vendors come in packages. The newer systems, particularly those merchandised by interconnect vendors, may be in module form. If you grow, you can plug in another module. But this may not be the case in Western Electric equipment or the equipment supplied by the independent utility. It is highly embarrassing to the telecommunications coordinator if he finds he has outgrown his system during the second year of a five-year contract. In such a case the old system must be removed and a larger system installed. The "termination charges" specified in the three- or five-year contract must be paid unless the consumer has exceptional negotiating skill.

The following discussion will give you the capacities and features of the series pricing systems except for Centrex, which will be covered thoroughly in Chapter 9.

## SERIES 100

The workhorse system of the telephone industry, a 701-type step-by-step dial system (Figures 1 and 2), is classified as Series 100. Such systems have a capacity of as many as 10,000 lines with a capacity for an appropriate number of central office and dedicated trunks. Other Bell System packages of dial equipment may also be used to serve Series 100. See Table 1, p. 101. The features of Series 100 were named in a Bell System presentation before a 1972 meeting, and I am using that official document in writing this portion of this chapter.

### Series 100 Features

1. *Attendant position.* In Bell System terminology, the "attendant" is the switchboard operator. Her "position" is the switchboard or console.
2. *Direct Outward Dialing (DOD).* This means that the user can secure an outdial line by dialing a single digit.
3. *Station-to-station calling.* This sounds as if it refers to the

**Figure 1.** Contrast in size between a step-by-step switch, which frequently serves Series 100 systems, and ESS equipment, which is used for the newer switching systems manufactured by Western Electric. *(AT&T Co.)*

**Figure 2.** 701 switching equipment. It is usually rented to customers whose station requirements are 80 or over. This switching equipment is a mechanical step-by-step means of completing all types of telephone calls. *(AT&T Co.)*

toll calls we discussed in Chapter 1. Instead, it means that a man sitting at one desk can call a man at another desk by dialing two, three, or four digits, depending on the size of the dial system.

4. *Call transfer—attendant.* If you wish to transfer a call which comes to you from outside your dial system, you slowly depress the carriage button on your telephone instrument. If you have a modern dial switchboard, a single touch of the button is all that is required. If you are using one of the older boards, you may have to depress it slowly several times until your operator notices the flashing of the supervisory lamp which appears at the position where your call was connected to your line. She then transfers your call.

5. *Restriction from outgoing calls.* Your dial system can be wired so that a person dialing for an outside line receives a busy signal.

6. *Station hunting.* Telephone stations are arranged in groups of 10 stations in the step-by-step switching system. Within any group

of 10 numbers "hunting" can be arranged. Let's say that your station number is 311. If you are talking on that line and someone within your dial system dials your number, the calling person receives a busy signal. Hunting permits the second call to be connected automatically *to the next sequential number.* Thus, calls to line 311 automatically rotate to 312, your secretary. If another secretary is assigned line 313, this sequence gives your further protection if your secretary is also talking on her line when the second call is received. In a customer service department, the lines may rotate throughout the series of 10 numbers until a free line is found. However, be careful about your arrangements on rotation of the line ending in the digit "9." This number rotates *backward.* Thus, line 319 will hunt to line 310, not line 320.

The hunting feature may be installed on dial systems which were not arranged for hunting originally. In such a case you change the telephone numbers of your people until they are in a correct relationship for the hunting sequence.

Let me caution you about arranging too much hunting. The hunted numbers should always include people who are competent and willing to help the person calling. It is better for the caller to hear a busy signal than to experience an "I don't care" type of response from a disinterested clerk in a nonrelated department.

There is no monthly rental charge for the hunting feature. A one-time wiring charge usually applies when you change or originate the hunting sequence.

7. *Night service.* What happens to your switchboard service when your operator goes home? Any switchboard or console can be arranged for night service. In the case of a cord switchboard, one end of a cord pair is placed in an incoming trunk jack. The other end of the cord pair is placed in a station jack. When that trunk number is dialed by someone outside the company, the station to which the cord is connected rings. Thus, that individual has his own private night number. However, if he should be talking on his line, the hunting equipment which controls your incoming central office trunks will automatically divert the incoming call to the next "night line," which may be in a far distant office.

If many incoming calls are expected to a specific night line, telco can supply special night numbers which do not hunt. Thus, the second incoming call would get a busy signal. This is usually an extra-charge service. When a console is used as a switchboard, night connections are usually preset and are put into effect by turning a key on the surface of the console.

Please see Chapter 11 for a description of night answering service. This permits anyone to pick up an incoming call from a guard station or other point which is manned at night.

8. *Power failure transfer.* Telco provides a limited number of lines which are automatically connected to its central office if commercial power should fail. Thus, a dial tone is furnished to those lines from the central office rather than from the dial equipment located in your office. An extra-cost alternative to this feature is called "reserve battery power." Auxiliary batteries located on your premises furnish sufficient power to operate the entire system for a limited period. When power fails, the automatic equipment switches the system to battery operation.

## SERIES 200

This equipment offering includes all of the features you have just read for Series 100. However, in some Bell System companies, it is offered only in a 200 station package with an appropriate number of trunks. Other Bell System companies are furnishing newer Western Electric equipment (Figure 3) or, in some cases, equipment manufactured by both domestic and foreign interconnect manufacturers. If the latter is true in your vicinity, your Series 200 package will have a capacity of 400 stations and appropriate trunks.

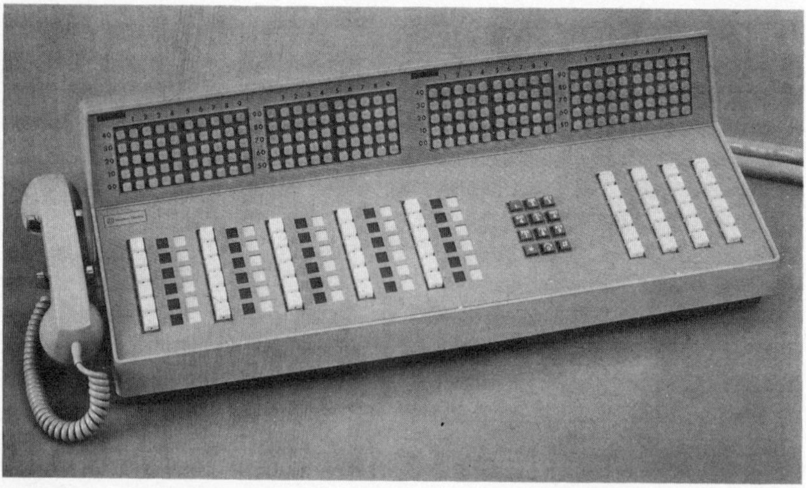

**Figure 3.** A new-style Western Electric PABX attendant console switchboard. This console is used with 757A or 770A equipment, which frequently serves the package dial systems offered in the Bell System's Series 200 and 300. The Direct Station Selection button panel is incorporated in the console itself. *(AT&T Co.)*

## Series 200 Features

9. *Attendant camp-on.* If you are talking on your line and an additional call from outside your dial system is received, your operator can depress a button which will "camp on" that call to your line. When you hang up, your telephone will ring automatically without operator attention.

10. *Indication of camp-on.* When the operator depresses her camp-on button, you will hear a faint beep. You now know that a second call is waiting. Your outside caller should not hear the beep if the system has been wired correctly.

We think that camp-on has two potential disadvantages. First, your operator may forget to give your second caller progress reports so that he knows your line is still busy. If you are talking on a long call, the second caller deserves those periodic progress reports and an offer from your operator or your secretary to arrange a conversation with someone else or a call-back. Thus, the camp-on feature requires periodic observation of the switchboard operation to make sure that your callers are remembered, not forgotten. Second, your train of thought is interrupted when you hear the beep. If you are in the midst of arranging the sale of a large quantity of merchandise, it may be better to avoid disturbing you so you can negotiate carefully without being hurried by the knowledge that an unknown second caller is waiting.

11. *Attendant Direct Station Selection panel with busy lamp field.* This feature is contained in a box which displays buttons. It stands next to the console itself, or it may be incorporated in the console. A recent Bell System offering, the 50 Series, provides a lamp display which is an integral part of the console instead of a separate busy lamp field. Each numbered line has an appropriate button. When an incoming call is received, the operator glances at the button which handles that line. If the line is busy, the button will be illuminated. If not, the operator presses the button and the call is automatically connected, ringing the called telephone. This arrangement is somewhat faster than connecting a call by using the console's dial. It may or may not be faster than a cord switchboard operation, depending on the skill and dexterity of the cord switchboard operator.

12. *Attendant conference.* A special piece of equipment is built into the console so that the operator can arrange a conference. The conference can be set up on an incoming call from the outside, or your people can call the operator and suggest that she arrange to have several of your people talk on an outgoing call which your

operator dials from her console. A conference circuit can be added to a Series 100 cord switchboard at a small additional monthly rental cost.

## SERIES 300

Series 300 has the same capacity as Series 200. It includes the 12 features which you have just read. In addition, it offers some very important features that resemble some of those which are standard in all the interconnected systems (see Chapter 12). Thus, be sure you understand the operation of the Series 300 features, as your telco marketing representative may be knocking on your door to present Series 300 as telco's answer to its interconnect competitors. If you wish Series 300 features and you need more than 400 lines, the Bell System will offer you Centrex II and possibly Centrex III. These telephone systems are described in Chapter 9.

### Series 300 Features

13. *Call transfer—individual.* In Series 100 and 200, the switchboard or console operator must handle the mechanics of transferring calls. This takes time, and when an organization has many calls which must be transferred the resulting work load may interfere with the prompt handling of incoming calls. In Series 300, each station line can be "split" by touching the carriage button of the telephone—providing the call is received from the outside. The caller is automatically placed on "hold" and the station user hears a new dial tone. He can now dial any station number *inside* the dial system. When the called telephone is answered, he asks if he may transfer the call. If the called person can help the original caller, the first person called touches his carriage button again and reports to his outside caller that he is transferring the call to Mr. Rhodes. He then hangs up and the outside caller finds himself talking to the person dialed by the recipient of the original call. All this is done without operator assistance and without the use of either hold buttons or line buttons.

14. *Consultation-hold.* In this case the person receiving the original call does not need to transfer the call but must secure needed information from Mr. Rhodes. He follows a similar procedure, but Mr. Rhodes hangs up when he has given the information. The original recipient than touches his carriage button and his original caller is brought back from "hold" to receive his information.

15. *Add-on conference.* The same procedure is followed. However, the person who was called for information stays on the line.

The original call recipient touches his carriage button, which brings the original caller back on the line. Now all three individuals can talk together.

16. *Trunk answer—any station.* This Series 300 feature provides night service. However, the user must ask telco to wire visible and/or audible signals in portions of the building where people are likely to work after the switchboard closes. When a call is received on an incoming trunk, the audible signal alerts those working. Any person can dial a code on his telephone which will automatically connect to the incoming call. Then, using the call-transfer mechanism, the call can be sent to any other office location.

Series 300 features definitely save the expense of multibutton instruments otherwise used by nonsecretarial personnel. If each executive telephone incorporates this "split-line" feature, why does an executive need anything more than an ordinary telephone without buttons?

The executive may feel that the removal of his flashing lights and buttons diminishes his prestige. The administrator must secure backing from top management if this saving in key system expense is to be implemented. Also, if the executive has to carry on lengthy conversations while he is getting information from a third party, it is best to supply him with the usual instrument with hold button and relay for his own line and for his shared line which is used for outgoing calls. This is because the Series 300 system provides only limited "consultation-hold" equipment. Too much use of the feature or long conversations may interfere with system operation.

## Supplying Series 300 Features for Larger Systems

If the features of Series 300 appeal to you and if you have a requirement for more than 400 station lines, you should ask telco's marketing representative about the availability of Western Electric's new 812A switch (see Figure 2 of Chapter 10). This unit is cabinetized; therefore no equipment room is needed. It is an electronically controlled crossbar system. Crossbar telephone equipment is dependable and takes less space and maintenance than step-by-step equipment of the traditional type. The 812A will serve up to 2,000 lines. This switch is also employed as the "machine" or serving vehicle on some Centrex I and II systems. Centrex II includes Series 300 features as standard offerings. Centrex has unlimited station capacity. You will find Centrex described in detail in Chapter 9.

## PRICING CONSIDERATIONS

In a very few Bell System companies unit pricing and series pricing are offered in the same tariff. You may wish to check your tariff and see if you have the option of selecting the pricing system which is best for you. You may likewise find that unit pricing for key systems is offered side by side with package pricing. Such a tariff indeed helps you, as you can pick the pricing system which gives you the lowest cost. Otherwise, you must insist that your telco marketing representative present an assortment of alternatives rather than a single plan. As features are added to the series packages, the rental rate per station climbs dramatically. You may elect to stick with the workhorse, Series 100, when you see your price quotations.

## SERVING MECHANISMS FOR PABX DIAL SYSTEMS

The telephone utility must supply the "machine" or "switch" which will make your PABX (dial system) work. If telco doesn't have a machine available in its reconditioned stock or in its new stock, it may have to order the equipment from Western Electric or an interconnect manufacturer. In any event, it is using part of its capital assets to provide the equipment to serve you, and it reserves the choice of machine to itself as long as the selected machine will provide the features you have ordered.

If you are in a territory where dial systems are priced on a unit basis, you may find systems offered which serve a stated number of lines and trunks. Rather than being series packages, these are packages of dial equipment which bear stated rental rates for the services furnished. Most of them are step-by-step systems or crossbar systems. A few packages provide features available only in an ESS. Some offer the features of the Series 200 and 300 systems.

Your tariff will list what is available to you. You may find the rental rates quite attractive for serving your needs. However, with the exception of one newer system, the 770A, the systems are not expansible. Be sure you make every effort to correctly chart the growth of your organization before you order.

When you are shopping for a telephone system, it may help you to know the capacities expressed in trunks and stations of the various dial packages which the Bell System uses in its sister companies. Not all companies offer every package. If a particular company has adopted series pricing, it will not necessarily reveal the identity of the serving mechanism to the consumer. But it may help you in your negotiations if you use a little "telephone language" and say to your marketing man: "Hey, Joe, how about finding me a 756A for that new system in our branch office?" The switch numbers plus trunk and station capacities are shown in Table 1.

**TABLE 1 Capacities of Trunks and Stations for Various PABX Systems**

| Bell System designation | Type of system | Line capacity | Trunk capacity |
|---|---|---|---|
| 805A | Electronic | 57 | 12 |
| 756A | Crossbar | 60 | 10 |
| 800A | Electronic | 80 | 20 |
| 757A | Crossbar | 200 | 43 |
| 801A | Electronic | 270 | 40 |
| 770A | Crossbar | 400 | 60 |
| 812A | Crossbar | 2,000 | 600 |
| 101ESS | Electronic | 4,000 | 384 |
| 701B | Step-by-step | 10,000 | As needed |

By the time you read this text, it is likely that this list of equipment will be much different. Remember that the *tariff* is the document which lists the telephone systems available in your vicinity. It is available to you in the principal business office of your telephone utility (see Chapter 6).

## 711 DIAL SATELLITE SYSTEMS

Up to this point, we have been discussing telephone systems which require a switchboard or console for operation. The Bell System has assigned the identification number "711" to a package of dial equipment which operates without a switchboard. Let me illustrate how the satellite system works.

Suppose that you have a research and development (R&D) laboratory which is located in a suburb 10 miles away from your principal location. Approximately 200 employees are located there, and most of them require telephone service. However, their need for service is principally between themselves at the suburban location and between that location and the headquarters location. The employees need to make occasional outside calls, and there must be a facility for receiving incoming calls from other locations of your company.

You see no need for a switchboard operator at the suburban location. The occasional visitors can be received by one of your accounting clerks who is located adjacent to the reception lobby. However, there are too many people at the R&D laboratory to justify the use of OPX (see Chapter 5) lines from your headquarters dial system.

Thus, you decide to install a 711 satellite dial system. Telco provides a package of dial switching equipment. It may be a step-by-step system or a crossbar system. It may come in a cabinet or you may have to furnish an equipment room that is properly ventilated, is clean, and has heat control. The system allows your research people to talk to each other by dialing two, three, or four digits. If they wish to reach headquarters personnel, they dial an additional digit, like "7," and then the three- or four-digit headquarter's station number. The satellite can be equipped with outdial trunks so they can dial "9" to secure an outside line. If their need for local calling and toll calling is minimal, they can dial "7," then "9," and use headquarters outdial trunks for those calls. This will save trunk rental charges for the satellite, an important cost reduction if you are located in an area of the country where flat-rate local trunks cost upwards of $30 monthly.

Headquarters people wishing to talk to the R&D people also dial a single digit, then the two- or three-digit number of the wanted person. Calls from outside the company come to the headquarters switchboard. The operator connects the call to a tie line connecting the two locations. She then dials the station number for the outside caller. Intraoffice calls are dialed directly over the tie lines without operator attention.

Many organizations use satellite telephone systems for plants and offices located in the same general area. The satellite system eliminates the rental of a switchboard and its associated usage contract. Salary and fringe costs are eliminated for the subsidiary location switchboard operator. From these savings must be deducted the cost for the needed tie lines, and you must also consider the additional burden on your headquarters operator, who will be receiving calls for R&D and dialing them over the tie lines. Your telco marketing representative can give you details of satellite costs, and his traffic department can work with you on the planning of the tie line and outdial trunk facilities. Be conservative. It is better to start with fewer tie lines and add to these facilities as traffic increases. Tandem dial repeating tie lines which probably would be required for a facility of 200 employees are contractual items which usually require five-year contracts for the needed switching equipment.

In considering satellite systems, you should also ask a qualified vendor of interconnected telephone equipment to quote on your needs. Many companies hesitate to install a major interconnected system without a test installation of such equipment in a location requiring a smaller system. The purchase or rental of such a system, assuming that you expect to remain in the subsidiary location for a reasonable period of time, may be quite comparable in cost to the satellite system, even considering the salary of the switchboard operator. As you will learn in Chapter 12 on interconnect, the newer systems sometimes produce handsome returns on investment in comparison to systems rented from the telephone utilities.

## COMPARING COSTS OF YOUR ALTERNATIVES

In this chapter, I have urged you to secure quotations from your utility salesman and interconnect salesmen on the alternative systems available to you. However, it may not be proper to compare the cost of your *existing* system with the alternatives quoted by your vendors and the utility. This is because your existing system may contain features which are unused or which are needlessly expensive in comparison to alternative instruments, features, or circuitry which may be available to you. This book has been designed to give you practical, tested ways to reduce the cost of equipment rental. May I suggest that you delay making your economic studies of alternative costs until you are sure that you've redesigned your *present* system for maximum economy? Chapters 10 and 11 are particularly important if you want to really reduce your rental costs. Then, when you are confident your system cost has been reduced as much as possible, you have a valid basis for comparing your current cost with the alternative costs quoted to you.

CHAPTER NINE

# CENTREX TELEPHONE SYSTEMS

Why do you hear so much about Centrex? What does it do for you and your customers? Will it really improve service and save money for your organization?

Basically, Centrex provides Direct Inward Dialing (DID). You have a specific telephone number which can be dialed directly by your caller. Thus, call completion is faster because the incoming call does not have to be forwarded to you by your switchboard operator.

However, there are many misconceptions about Centrex. Further, recent developments in technology are producing interconnected telephone systems which save operator labor, provide Centrex-type features, and still retain the advantages of a single telephone number.

The telephone utilities usually offer a type of Centrex which provides housing for the equipment in one of the central offices of the telco. It is difficult for a competing vendor of an interconnected telephone system to secure traffic information if the machinery serving the customer is located on telco premises. Thus, it may be difficult to make an economic study on the possible replacement of such a Centrex system with an interconnected system. It is only natural for the utilities to market Centrex aggressively, as such marketing will help protect equipment rental revenue. Once the Centrex decision is made, the Centrex user will hesitate to change his telephone system within the term of the typical three- to five-year Centrex contract. Thus, it is important that you learn the characteristics, advantages, and disadvantages of Centrex before you sign a Centrex contract.

## CENTREX AVAILABILITY

Telco cannot be *required* to offer you Centrex. This is only fair, because the capital investment involved in its acquiring Centrex equipment may be very large. All telephone utilities must necessarily think carefully about the cash outlays needed to expand their physical plant. In some localities the borrowing power of the utility may not justify expenditures for Centrex equipment. In many vicinities Centrex is sold and placed under contract before the serving equipment is ordered. This protects the utility's ability to cover its financing outlay with income from Centrex rental.

## CENTREX HOUSING

Telco must make another decision: Shall equipment be housed on your premises ("CU" Centrex) or on their premises ("CO" Centrex)? If you are willing to supply housing space, telco will reduce the monthly rental cost slightly for each Centrex line, as telco may save the cost of making an addition to its central office facilities. However, the decision to offer Centrex "CO" or "CU" is telco's.

## CENTREX SYSTEMS

Centrex machinery is of several different types. Telco decides which machine will serve the Centrex system you choose. Your machine may be step-by-step switching equipment similar to that which serves most older dial systems. If you require more sophisticated features, telco may install a modern crossbar or electronic machine. The most common Centrex systems are Centrex I and Centrex II. In some areas the Centrex III system may be offered.

### Features of Centrex I

Essentially, Centrex I provides DID. If you can educate your callers to learn your individual Centrex number, your calls come direct to you without being handled by your switchboard operator.

Thus, call completion is faster, and you will need fewer switchboard operators.

If you wish to have an incoming call transferred, you press the carriage button on your instrument once. This calls your Centrex switchboard or console operator. You then give her transfer instructions.

Centrex I gives you automatic accounting for toll expense. You receive a separate toll bill for each numbered Centrex line. In most areas, that bill likewise totals local message unit charges incurred by the user of that particular line.

If you have branch locations within the same central office area in which your main office is located, Centrex I may serve those offices without your paying mileage charges for lines which run from your office to the branch offices. For instance, banks with many downtown branches find these savings important.

### Features of Centrex II

In addition to the features described for Centrex I, Centrex II offers the four features normally associated with the Bell System's Series 300 package dial systems. These features are also standard on all modern interconnected systems manufactured by independent vendors.

**Consultation-Hold.** This permits you to touch your carriage button once, which places your outside incoming call on "hold." A new dial tone is heard. Now you can dial any other telephone within the Centrex system. You consult with that person. When you have secured the needed information, you touch the carriage button again. This brings your original caller back to you. You can now complete your conversation.

**Add-on Conference.** This repeats the same procedure. However, the person you call to get information does not hang up; he stays on the line. When you touch your carriage button to bring your original caller back on the line, the three of you will be talking together.

**Call Transfer—Individual.** Again, the same procedure is repeated. But in this case you tell the person you dialed that you would like to transfer your caller to him. If he agrees to help your caller, you hang up and your caller is automatically transferred to the person you called. This is done without assistance from your switchboard or console operator.

**Trunk Answer—Any Station.** Calls can be answered after your switchboard has closed for the day. Audible signals are installed in areas of your office where people may be working after hours.

When the signal sounds, any person can dial a code and thus answer a call which is ringing on your main switchboard number.

Centrex II always costs more than Centrex I, but its additional cost is partially offset by savings in multibutton telephones. This is because each telephone has a "split-line" feature. You do not need a "hold" button, nor do you need one button for your line and another line button to use for securing information. Consequently, most executive phones can be plain phones without buttons. However, if the user frequently must have lengthy consultation-hold calls, he should have a button-equipped phone, as the split-line feature may have an automatic time-out. Also, consultation-hold may not be available at a particular instant if the Centrex system is overloaded with others using the same feature simultaneously. If a secretary must furnish backup service for answering phone lines other than her own, she will still require the usual multibutton instrument.

### Features of Centrex III

This form of Centrex is served by electronic switching equipment. Such machinery is very expensive. Consequently, many telcos have not yet offered Centrex III to their customers. Here are some typical features.

**Camp-on.** If a caller dials the number of the main Centrex console or switchboard, the operator can "camp on" that call to the line if she finds it is busy. The called person hears a beep, which indicates that another call is waiting. Depending on the wiring of the system, the calling person may or may not hear the beep. When the called person hangs up, his phone rings automatically, and he is connected to the waiting person. Only one waiting call can be camped on to a station line. A second waiting call must be handled by the Centrex switchboard operator in the usual way.

**Call Forwarding.** If you are going to another office, you dial a special code plus the number of the telephone where you can be reached. Your incoming calls are forwarded to that number automatically. When you return, you must dial another code to erase your original instructions.

**Compressed Dialing.** The machine gives you access to a list of frequently called numbers. After you dial a few digits, the machine will dial all the digits necessary to complete that call.

**Group Answer from Any Station.** In a particular department with related phones, the department secretary can be equipped with an audible signal which indicates that someone's telephone is ringing. By dialing a code she can answer that telephone. This feature may eliminate both multibutton phones and the expensive relays which provide the flashing lamps found on multibutton instruments.

**Touch-Tone Calling.** This feature is standard in Centrex III installation; it may be available as standard or as an extra-cost feature on Centrex II systems, depending on telco's central office facilities. Touch-Tone calling may also be available as an extra-cost option on Centrex I systems.

**Switchboard Transfer.** If you do not answer your telephone and if your secretary does not answer, the incoming call may be routed to the console or switchboard operator. Similarly, if lines are busy throughout whatever rotation sequence has been arranged, the operator receives the call. This feature may also be offered as an option on Centrex II systems.

## CENTREX INSTRUMENTATION

A *primary* Centrex line has its own number. It can make or receive any type of call unless it carries some kind of restriction on outgoing calls, such as restriction from the ability to dial toll calls or multiple message unit calls.

An *extension* Centrex instrument is a "bridged" telephone. It shares another person's Centrex line, so it costs much less. When a number of persons in a department require telephones but have few incoming or outgoing calls, it may be best to equip them with extension Centrex lines. A departmental secretary answers all incoming calls on comparatively few primary Centrex lines. She then distributes calls to those using both primary and extension Centrex instruments by using a dial or voice intercom system.

A *fully restricted* Centrex line is for conversation within the Centrex system. It has its own number. It can call any numbered line in the system or it can be called by others. However, it cannot reach the Centrex or telco operator, nor can it receive or make outside calls. Usually it is much less expensive than a primary Centrex line. Almost every organization can identify clerical jobs which do not require outside dialing or outside call-receiving capability. Thus a fully restricted Centrex line will reduce equipment rental cost substantially for such clerical positions. Such a line also prevents the user from incurring any cost for personal toll or message unit calls. We find that many present Centrex users do not know about the availability of fully restricted service, although it appears in all Centrex tariffs.

## CENTREX CALL RESTRICTION

I have described the fully restricted line. Other classes of restriction are available. A line may dial local calls but not toll calls. Another line may receive incoming calls but not have a local or toll-dialing capability. Another line may dial single message unit calls but not multiple message unit calls. Still another line may dial anywhere in the message unit area but may not dial toll calls. Your telco representative can give you information on the restriction capabilities of the system offered you.

## CENTREX CALL ROTATION FOR BUSY LINES

If your Centrex system is to be served by step-by-step switching equipment, call rotation (hunting) will be similar to that of the conventional dial system. A 10-number group is involved. An incoming call encountering a busy line can automatically hunt to the next sequential number in a group of 10 numbers. However, the final digit "9" switches to the preceding "0." Example: Line 3129 will switch to line 3120, not to line 3130.

If your system is served by crossbar or electronic equipment, hunting is much more flexible. Lines can be arranged to hunt backward as well as forward, and numbers can be skipped so that hunting may be done anywhere in a group of 100 numbers or, in some systems, within a group of 1,000 numbers. Your telco marketing representative will give you full information on the possibilities inherent in the system being offered you. Hunting is then wired to your specifications.

## INTERCEPTION PROCEDURES

If you've done the educational work correctly, and the postcards showing your DID number have been mailed on schedule, your "cut-over" to Centrex may occur without trauma. But . . . remember that many people will be calling your old number. If you tell your telco to put the old number on their intercepting operator service, that operator will be giving your main Centrex switchboard number. Then, your overloaded console operator will have to educate your callers and give them the new number. For the first few weeks you will require some additional consoles or switchboards to handle the extra load.

Progressive telcos will offer to keep your old switchboards

operating on a no-charge basis. Those calling your old number on toll or from coin stations will not incur a toll charge or lose their dime when your *own* operator answers. Your operator can give your caller the specific number he should dial to reach his party directly. Thus, through the courtesy of your telco, you continue with your educational work. And in some cases of exceptionally cooperative telcos, you may receive an offer for installing tie lines between the old system and the new system. In this arrangement, your customer does not have to redial. Your operator on the old switchboard can connect your caller to the new Centrex system via tie line, dialing the new number for him. If you are about to adopt Centrex, *be sure* to keep your old switchboard in operation for a few weeks so that your expert operators can give correct numbers to your callers.

## CENTREX SWITCHBOARDS

Centrex switchboards are available either in console or in cord switchboard form. The cost is usually the same. Most telcos seem to recommend the console. However, if there is heavy usage of operator-controlled WATS, tie lines, or FX circuits, the cord switchboard may provide far better service to users. This is because those lines can be extended directly to users by the operator without requiring the user to disconnect so the operator can call back to extend the special trunk.

The cord switchboard likewise provides visual indication of calls waiting on unanswered trunks. This factor stimulates operators to work faster during seasonal or daily peaks. On the Centrex console there is a call-waiting lamp, but the operator does not know *how many* calls are waiting.

## DIRECTORY LISTINGS FOR CENTREX

If an executive has his own DID line, it is only natural to expect that his number will be listed in the white pages of the city directory. In some cities, banks adopting Centrex list all officers. This is very expensive. A Centrex system includes only one "free" directory listing—that of the main number of the switchboard. Supplementary listings are billed at monthly costs running from 50 cents upward. Also, the federal excise tax applies to supplementary directory listings.

We suggest that you study the quantity of incoming calls for each department which has public contact. Then, list those departments in alphabetical order *first*. At the end of the listing section, have a line which says, "For all other departments or individuals, call XXX-XXXX." We feel it is a mistake to list the main switchboard number at the top. Many people dial the first number they see. In Centrex, we want to do our best to make sure that calls are indeed dialed directly to the department or individual.

## EMPLOYEE-ANSWER RESPONSES FOR CENTREX

Remember: every employee answering a Centrex telephone is a switchboard operator. In areas where the public calls, we suggest that the answering person identify the organization first, then give the department or individual's name. Example: the First National Bank of Dallas carefully trains all telephone-answering people to say "First in Dallas" as an opening phrase. You can readily see the marketing implications, in addition to the good manners of letting the caller know he has reached the right destination.

It is equally important to continuously monitor the answer response used by your switchboard operators. We suggest a phrase like this: "I'll be glad to connect you, sir. Wouldn't you like to have Mr. Brown's direct dialing number so you can reach him more quickly?" In some companies, the operator says "You may reach Mr. Brown directly by dialing XXX-XXXX" and then pauses. The customer thinks he must make a second call, and if he is at a coin station, he reacts poorly to the suggestion.

Centrex has been installed widely in educational institutions. Because the student body changes constantly, the main Centrex consoles are overloaded with calls from parents and students asking for the numbers of dormitory Centrex telephones. To lessen this problem, the institution can establish an off-switchboard directory information position, perhaps manned by inexpensive student labor. The consoles are kept free for calls from outside. Likewise, the Centrex console operator will be able to take time to connect the caller to the desired number rather than ask him to make a second call.

Here's one more suggestion. When a caller reaches the main switchboard, if the operator, after making the correct response, finds the requested line is busy, she should say: "I'm *so* sorry, sir, but that line is busy. If you'll call back in a few minutes, you'll be able to reach Mr. Brown directly by dialing XXX-XXXX." This procedure clears central Centrex consoles of those calls and frees circuits for people who are waiting for an incoming line to your main number.

## CENTREX ADVANTAGES AND DISADVANTAGES

Because Centrex systems are being marketed actively by the utilities, the prospective Centrex customer must carefully analyze the claimed Centrex advantages and be conscious of some important disadvantages. Let's discuss these concepts.

### Direct Inward Dialing

If speed of handling incoming calls is critical, DID may indeed be helpful. But *caller* education is required. Do the *same* people call you? Or do you receive only occasional calls from a large group of people? In general, a manufacturing organization or bank can do a far better educational job than can a newspaper, department store, or hospital.

### Savings in Operator Labor

In a well-planned Centrex installation, 75 to 85 percent of incoming calls may be dialed directly to the station number. Fewer switchboard operators are required. But you must be realistic about the need for providing answering service for ringing telephones which are directly dialed. This means that you must provide one or more additional multibutton answering points for any line which can be accessed by important people. You cannot afford to offend a customer by letting a telephone ring indefinitely. Thus, you are actually moving operator labor out of your switchboard room and applying it on multibutton sets located within your departments. Those instruments *must* be manned throughout the working day, so another person must move into that position during lunch hours and relief periods. In your economic study, be sure to calculate the extra cost of departmental labor, multibutton instruments, and flashing relays when you are figuring Centrex costs.

### Station Billing for Toll and Message Units

Manufacturers of interconnected telephone systems are offering automatic toll-ticketing devices as optional features of their equipment. A prospective Centrex user who is adopting Centrex chiefly for accounting purposes should talk to the marketing representative of the independent vendors.

## The "One-Number" Concept

Many sales executives want to promote the company's main telephone number as a point of contact where skilled operators will direct the customer's call to a department which will give him proper attention. In Centrex, each answering person becomes the voice of the company. Careful training must be done on the use of a proper answering response. The customer must know that he has reached the right destination. If the call has hunted to a line other than the intended line, the answering employee must be carefully trained to see that courteous, expedited service is given. Wrong numbers must be handled with special care. It is evident to most public relations officers that it is far easier to train a small switchboard staff in careful call handling than to train each person answering a Centrex line, particularly when the organization may have high clerical employment turnover.

A telephone switchboard operator's primary responsibility is prompt answering of incoming calls. Once the call is connected to the requested line, her next most important responsibility is to make sure that the called telephone answers. If it does not, or if the line tests busy, she takes the responsibility for finding someone who can help the waiting caller. In Centrex systems, poorly trained clerks or secretaries may place calls on "hold" when lines are busy. The caller may be forgotten or ignored for lengthy periods. Also, departmental personnel cannot be expected to have wide knowledge of all company functions. As a result, calls may be subject to multiple transfers before help is found. For these several reasons, we believe that organizations sensitive to building good public relations should seriously question whether Centrex will give a proper grade of telephone service to their callers.

## Caller Education

In many companies, executives frequently move from one department to another when their responsibilities change. The question arises: Shall we permit the executive to take his Centrex number with him? This action may interfere with proper call rotation on busy lines in the department he is leaving. On the other hand, if he is assigned a new number, those calling him must learn the new number if Centrex is to be effective. When the Centrex system is first installed, as well as when people move, education will require new business cards, stationery, and forms. Announcements must be mailed on a careful schedule. Company advertising must deemphasize the main Centrex number. All this effort is costly, and costs will continue to mount as people move.

### Federal Excise Tax

The percentage amount of this tax is being reduced gradually. Further, owing to an IRS administrative ruling of some years ago, only the portion of Centrex line cost designated as "Exchange Access" is subject to tax; the portion designated as "Intercommunication" within the Centrex system is no longer subject to tax (see Chapter 6). However, *every* Centrex primary line is subject to federal excise tax, even under the new rules—and this cost normally is greater than the tax applicable to the central office trunks which serve the traditional dial or interconnected telephone system. Be sure to calculate this tax and add it to your Centrex cost.

### User and Customer Reaction to Centrex

If these problems exist, why do so many major firms use Centrex? Are they happy with these systems? Do callers like Centrex?

We have found that most users have bought the Centrex concept because skillful marketing was done by the telephone utility. We find that many cost factors may have been omitted from the economic study which should be made before the Centrex decision. When the user has allowed key system expense to grow uncontrolled, the features of Centrex II may reduce that expense by a certain percentage. However, Centrex systems are almost always more expensive than the system they replace, sometimes by as much as 30 percent. Rigorous attention to the design of key systems (see Chapter 11) can reduce this difference, but if the same attention is paid to key system design in an existing system, Centrex cost differentials will be greater.

Caller reaction to Centrex frequently is highly critical. Unanswered telephones, busy lines, and slipshod handling of calls by untrained clerical personnel all detract from the corporate image. In most organizations, we think that the backup service provided by a carefully trained switchboard operator is indeed superior to dividing this responsibility into departmental segments. True, Centrex options may provide for unanswered and busy number calls to be diverted back to the Centrex operator. But those options are not selective—they must apply to every numbered line. Their cost is substantial, and the option requires that the user add the needed operator labor, thus eliminating one of the important advantages claimed for Centrex.

## Current Technology

There is another development which any prospective Centrex user should carefully consider. This is the speed of the consoles used by operators in the modern telephone systems marketed by independent vendors and interconnected to the utility network. Because of the rapid action of the call-forwarding mechanism on the newer consoles, we believe that the new equipment may cut operator-handling time by as much as 75 percent. Thus, savings in operator labor can be accomplished without sacrificing the important advantage of skilled interrogation of callers and careful backup on unanswered or busy lines. On some of these newer consoles, if a line remains unanswered, or if an incoming call is "camped on" to a busy line for more than a fixed number of seconds, the operator is alerted by that line number being flashed on a display panel. Thus, she is reminded to give a progress report or to take action to help the caller.

Centrex will continue to be a major offering of the utilities, but we expect that fewer users will convert to Centrex when they learn the advantages of the alternative systems available. Also, a number of companies and banks that use Centrex are considering returning to the "one-number" concept in order to maximize service to callers, particularly when the modern interconnected systems save operator labor, improve public relations, and at the same time provide a handsome cost reduction over the cost of Centrex service.

# CHAPTER TEN

# SUGGESTIONS FOR IMPROVING YOUR TELEPHONE SYSTEM'S PERFORMANCE

This chapter is designed to assist readers who have not been trained in telephone traffic engineering techniques. The suggestions which follow utilize do-it-yourself methods. They also encourage you to take advantage of the many free services of your telephone utility.

The traffic engineers associated with telephone utilities and interconnect vendors are dedicated people who genuinely wish to equip you with adequate facilities which will properly handle the traffic volume generated in the busiest portions of your working day. However, their workload is heavy, and they may not be able to review your telephone service regularly.

Also, there is a possibility that your business activity may have changed since you installed your telephone system. You may have too few or too many trunks or switches for your current business volume. Your operators might be able to increase their productivity if they could learn new techniques of operating their switchboards or consoles. By using some easily learned techniques, you may be able to spot potential trouble in your switching apparatus before complaints from your users begin to reach you. This chapter describes the action you can take to ensure that your telephone system performs at peak efficiency.

## BLOCKED-CALL STUDIES CONDUCTED BY TELCO

A "blocked call" is one which encounters a busy signal when the number is dialed. It is important that your callers reach your main telephone number with a minimum of "busies." If you have heard complaints that your lines are busy or if your switchboard operator reports that all incoming trunks are filled at frequent intervals, you should ask your telco marketing representative to arrange for a "busy study," that is, a blocked-call study. This work is done in telco's central office. Meters are connected to your central office incoming trunks. You should order the study made on days when you expect traffic to be heavy. Telco will then report the number of busies recorded on each day studied. If possible, ask telco to give you their report by hour so you can chart the busies throughout the working day.

Busies relate to a technical term known as "grade of service." In turn, grade of service is related to a standard of service expressed as a "PO" factor. PO is an abbreviation for Poisson, a man who developed the mathematics behind certain tables used universally in the telephone industry to determine the number of trunks which should be installed to give your callers a proper grade of service. In general, the Bell System will engineer your system for a PO-2 standard. This means that for every 100 attempts to dial your number, two busy signals will be recorded. In some organizations, private line or other dedicated facilities used only by company personnel are engineered for much higher factors, like PO-10 or PO-15. The higher the factor, the less trunks are required and the more the costs are reduced. Management may reason that it would rather have people keep dialing the dedicated facility than supply additional expensive tie lines or other dedicated facilities to reduce the busy signals.

However, the engineering of incoming central office trunk facilities requires a different approach. We are now dealing with your organization's public relations with its callers. Busy signals are very annoying. Thus, we agree that a PO-2 standard is a good standard of service and we recommend the installation of sufficient trunks to achieve that standard.

A busy study conducted only at the telco central office is not sufficient, however. It is essential that you know how many calls actually reached your switchboard on the same day that the study was made in the central office. When you know the *total* calls handled at the switchboard, you can then look at the busies reported and determine if you should rent more trunks. For instance, telco reports that 40 busy signals were heard by your callers on a Monday. That sounds as if you are giving a poor grade of service to those calling you. However, telco's service representative stood behind your operators on that Monday observing their call handling and tallying the calls received on your incoming trunks. Your operators handled 2,500 incoming calls. The 40 busy signals are well within the 2 percent busy allowance demanded by a PO-2 standard. You conclude that your present number of trunks is adequate for your current traffic.

Caution: don't dismiss this matter from your mind once you complete a study. Busy studies are a no-charge service of telco. You should order them made as often as your business volume shows major changes.

What should you do if the busy study shows zero busies or just a few busies? If the study was taken on a day which is one of the busiest of your week or month, this indicates you may be overtrunked. You can ask telco to disconnect a trunk or two. You then make another study. Trunks are noncontractual items and can be added or subtracted as required by traffic conditions.

Here's another situation. Your busy study from telco indicates that you have adequate trunks, but your customers still complain that they hear busy signals when they dial your number. Such complaints may pinpoint the fact that the central office of the telco serving you is overloaded. These situations are very common. The busy signal heard by your caller is not a "line busy" but is an "exchange busy." The latter signal usually sounds slightly different from the "line busy" signal. This situation calls for a candid discussion with your telco representatives. Possibly you should talk to one of the top telco officials, as your telco is required by your state's regulating commission to furnish a specified grade of service to its customers. If you are receiving less than a proper grade of service, you can place a formal complaint before your state commission if telco has not taken action to correct the problem.

In most situations of this kind, telco is quite aware of the difficulties. However, the modernization of a central office to relieve crowding takes many months and sometimes years. Your only current alternative may be a change of telephone number so that your telephone service can be furnished by another central office of adequate capacity. You'll have many problems in planning for such a change, but if your customers cannot reach you within a reasonable service standard, you should consider making the number change.

## BLOCKED-CALL STUDIES CONDUCTED BY YOUR SWITCHBOARD OPERATORS

The telco study discussed in the preceding section may take time, as other organizations may be competing for the metering equipment which must be used. If you believe you have a problem, ask your switchboard operators to help. At 10- or 15-minute intervals, they glance at the trunk jacks on the board and quickly count the number of cords which are inserted in the jacks. You can plot the resulting data on a graph and determine if you are reaching the capacity of your group of incoming trunks. Such an informal study also gives you immediate information on the possibility of overtrunking. Your operators can make similar studies for you on dedicated facilities controlled from the switchboard. If you have dial-access tie lines, your operators can observe their use at any one time by counting the number of illuminated lamps which indicate the line is in use. Such a study may help you increase or decrease tie lines or your dedicated facilities as your needs change.

## SOLVING PROBLEMS INVOLVING INCOMING SEQUENTIAL TRUNKS

A "trunk" is a telephone line which connects your switchboard with telco's central office. A PABX dial system requires enough incoming trunks to give a proper grade of telephone service to those who wish to reach you. It is customary to arrange incoming trunks sequentially, which means that if your main number is 498-3111 the second trunk will be 498-3112, and so on through the series. The wiring of your particular system may be arranged so that nonsequential numbers can be made part of your hunting sequence. Thus, if line 498-3113 is assigned to another subscriber, telco may give you an entirely different set of numbers. In such a situation, line 498-3112 might rotate to line 498-3121, with additional sequential numbers assigned from that number onward. It is important that calls dialed to you automatically hunt to an unoccupied trunk.

Now, suppose that you need additional incoming trunks. Your telco representative may tell you that your particular dial system requires all trunk numbers to be in an exact numeric sequence and that no further sequential trunks are available to you. The next numbers in your series already are being used by other customers of telco. This is a serious situation. Here are some suggestions on possible remedies.

### Change the Numbers of Telco's Other Customers

If the numbers have been given to residential customers, telco may be willing to talk with them and see if they are willing to take another number. However, if telco's customer is a business, most telcos will not ask the business to give up its number. In this situation, I suggest that you dial the number you would like to acquire, determine the identity of the business and its manager, and arrange an appointment to discuss the problem. If this matter is of real importance to you, a monetary consideration may be of interest to telco's other customer. Perhaps I should just be candid and say: "Make him a good cash offer to give up his number!"

### Consider "Grading" Your Incoming Trunks

This alternative involves the insertion of trunks in between the current sequential numbers. Thus, if one trunk is 371-4018, another trunk may be inserted which will be called 371-4018-A. Such graded trunks do not offer you the grade of service which is possible with additional sequential numbers, but telco should be questioned on whether trunk grading is possible with your particular dial system.

### Consider Using Nonsequential Trunks

Is your operator using sequential trunks for calls which she is dialing from the switchboard? Sometimes top executives like the prestige factor of calling the operator and having her dial local or toll numbers. That dialing is usually done on trunks which should be reserved for incoming calls. If you cannot discontinue this practice, consider the installation of nonsequential trunks on the switchboard. Then direct your operator to do her outdialing on those trunks.

### Consider Using Private Business Lines

If certain of your departments or individuals receive heavy incoming call traffic, you may wish to install private business lines which will be located in that department. The lines can be installed in hunting sequence for customer convenience. The numbers are issued to selected callers. They call directly, as they would in a Centrex installation. The switchboard load is reduced accordingly, thus postponing a possible change of your main telephone number.

### Consider Adding Outdial Trunks

In a later section of this chapter, I will discuss the several kinds of trunks available to you. In some dial systems telco has installed a type of trunk which can accept incoming calls from outside the company and can also accept overflow calls from those who dial "9."

If these trunks, which always appear on the switchboard, are part of your sequential group, you may wish to change them from "two-way combination" trunks to "two-way" trunks. You can then add additional one-way outdial trunks in your equipment to give adequate service to those dialing "9." This change will give better service to your incoming callers.

### Consider Changing Your Telephone Number

If none of the previous alternatives is satisfactory to you, then you can change your main telephone number, making sure that telco reserves sufficient numbers in the new exchange to take care of your estimated future growth.

## DETERMINING THE NUMBER OF SWITCHBOARD POSITIONS

Telco offers every subscriber the free services of skilled traffic observers. These observers visit your switchboard and watch your operators handle traffic, as described in the section "Switchboard Service Observing" in Chapter 17. They offer suggestions on better manual handling of the cords or console buttons. They tally and tabulate every operation performed. Then they apply a standard time factor to each separate operation. For instance, the telephone industry uses a standard switchboard labor factor for the time taken by the operator to place a cord in a trunk jack to answer an incoming call, speak to the caller, and connect that call to the appropriate station jack. By applying these standard time factors to all the work performed by the operator, the traffic observer can give management an hour-by-hour chart which shows the number of operators needed hourly to handle traffic received at the switchboard.

Through studying such information, you know if you should be thinking about installing another switchboard position so you can give a proper grade of service to your callers. If you have enough switchboard positions, but you need additional operators during the busy hours, perhaps you can advertise for skilled part-time help. Many ex-operators are now bringing up families. They might enjoy working during the busy hour of the morning, then returning to greet the children after school hours.

If telco suggests that you need an additional switchboard position, be sure to read the section in this chapter on console switchboards. Also, keep in mind that a switchboard acquisition is usually subject to a five-year contractual arrangement. If you should discontinue use of the switchboard before the end of the five years, you would pay a certain percentage of the agreed rental over the life of the contract.

Thus, before you order the additional switchboard position you may wish to think about the possibility of changing your entire system. Ask your telco marketing representative for his recommendations concerning crossbar and ESS equipment which would bring you modern technological advantages. Also ask your interconnect salesman to compare the cost of his offered systems with that of telco's recommended systems. At the present rate of technological advancement in the industry, be very cautious about binding yourself contractually to a unit of equipment which may be obsolete.

## DO-IT-YOURSELF TRAFFIC STUDIES

As telco's traffic studies must sometimes be scheduled many weeks in advance, you may wish to make your own traffic study. Visit your operators and watch the bright lamps which indicate that an incoming call is waiting to be answered. If the lamp is not extinguished within a few seconds, you are giving less than a proper service standard to your incoming callers. You may need additional operators at any unmanned switchboard position, or you may need an additional position. Or you may not need either—sometimes operators are performing call dialing or paperwork operations which prevent them from answering incoming calls promptly.

Switchboard observation is an important part of the telecommunications coordinator's responsibility as well as that of telco. In observing cord switchboard operation, a key question should be asked of the operator: "Do you ever run out of cords?" If she does, you have a problem on which you must take immediate action. Every switchboard position has a certain number of cord pairs. If traffic on the switchboard requires the use of all available cord pairs and one or more additional calls are received on the incoming trunks, your operator is helpless. She cannot answer the calls because she has no more cord pairs. Your caller wonders why you don't answer your telephone lines. An additional switchboard position must be ordered speedily in such situations. Until it is delivered, you may wish to install special incoming business private lines in departments with heavy incoming calls. You issue the special numbers to selected callers and those calls now bypass the switchboard, relieving the congestion there.

## ENSURING ADEQUATE SWITCHBOARD COVERAGE

As mentioned earlier, far too few top executives appreciate the critical importance of the switchboard operating function. Operators may be poorly paid. Absenteeism runs high. If sickness strikes the regular staff, possibly removing two operators on a single day, the company's operations may be severely hampered.

The telecommunications coordinator should keep in close touch with the switchboard supervisor or operators to ensure that the positions are adequately manned at the busy times of the day. Frequently we find that there are sufficient switchboard positions but that operators are not available to man the positions.

Any management should be sure that relief operators are available and on call. The alert management will select relief operators from employees who have general knowledge of company operations and who possess pleasant voices. These people should spend several hours a week at the switchboard to learn telephone numbers and keep up-to-date on personnel changes. In the beginning they are sent to telco's offices for specific training by experts. Then, when traffic gets beyond the handling ability of the operators on duty, the relief operator is summoned to help. Her department head must understand the importance of her work so that he will release her immediately for switchboard duty. By thinking ahead on the training of relief operators, the coordinator may save himself some unpleasant experiences, for example, when a traveling chief executive finds that he is unable to reach his own company by telephone.

## METER READING

Every dial system is equipped with meters which help the user to secure information on the adequacy of his facilities. In the Western Electric 608 dial switchboard the meters may be incorporated in one of the switchboard panels. In step-by-step equipment the meters usually are mounted on one of the racks in your equipment room or, in rare cases, in the casing of the switchboard. If you are using a package dial system enclosed in cabinets, ask your repair service technican to locate the meters and show them to you. In some telcos, meters are called "registers." You might use that term if your repairman doesn't quite understand what you wish to find.

The most helpful meters are those connected to your outdial trunks, tie lines, and any dedicated facilities which are installed in one or more of your dial levels. A "level" is a group of 100 numbers. For instance, if you dial a "7" to access the telephones at your research and development laboratory, you are using the seventh level. The outdial trunks are usually installed on the ninth level.

In the meter group, you'll notice some identification letters and numbers above the dials of each meter. LTB indicates a "Last Trunk Busy" meter. ATB stands for "All Trunks Busy." DCO means "Dial Central Office." Lev 9 means that the meter is associated with the ninth level. PC stands for "Peg Count." Such a meter moves every time its associated switches are engaged, so readings will show much more activity.

If the meter is ATB, it will advance a single digit when a person dials "9" and finds all the outdial trunks busy. If the meter is marked LTB, it will register when someone dials "9" and secures the last available outdial trunk. This does not necessarily mean that the next person dialing "9" will secure a busy signal, as meantime another outdial call may have been terminated.

The ATB and LTB meters associated with levels of your exchange help you adjust outdial trunks and dedicated facilities to the needs of your users. If you read the meters regularly and plot the reading on a spread sheet or graph, you will have a picture of this portion of your telephone traffic. Thus, you can make better decisions on adding additional trunks or tie lines. Conversely, if the meters show little movement or do not move at all, you are equipped with more trunks, tie lines, or switches than you need. Before you issue orders for additions or removals, be sure that all your trunks and switches are *working*. You are paying a monthly rental for trunks; you may also be paying a rental fee for each switch. Your repair technician can test all equipment to make sure it is functioning. He can also make a trunk "busy" before removing it, so you can double-check on your meter reading. Thus, you have taken a trunk out of service without actually removing it, and you can again check your meter readings to verify your initial conclusions.

You should read meters hourly if your weekly readings show a considerable movement. You can then relate your reading to the "busy hour" and make a better judgment on the grade of service you are giving those who are making outdial, tie-line, or other dial-access calls. It is possible that you can tolerate some delays during the busy hour if your trunk facilities show few LTB or ATB readings during the off-peak hours.

You'll find many more meters in your equipment. Some of them may not be connected. You can tell by looking at the reverse side of the meter and determining if wires are running to some group of switches. Meters are connected to each group of line-finder switches. Such an LTB meter tells you how often people engage the last line finder in that group of switches. If this meter shows a lot of movement, you may begin to hear complaints when your people receive no dial tone when they pick up their handset. Similarly, other meters relate to selectors and connectors which complete your intraoffice calls once you hear the dial tone. Sudden or unusual movements in any of these meters may denote trouble and you should immediately call your telco marketing representative. He will ask one of his traffic engineers to survey the situation.

Actually, the meters are installed not for your information but to give telco's traffic engineering department periodic information about the working of your telephone system. You may find that your operator has been instructed to read the meters regularly and send those readings to telco on special forms provided for that purpose. However, in common with most of us, telco's engineers have a heavy workload and it may be some time before your meter reading comes to their attention. That is why we suggest this do-it-yourself project on regular meter reading.

## REBALANCING TRAFFIC LOADS

Suppose that when you installed your telephone system you assigned your "heavy talkers" in the sales department and purchasing department to the 200–300 group of numbers in your dial equipment. Your "light-talking" engineering people were given numbers in the 400 group. Your business has been growing rapidly. Your purchasing and sales people are complaining that it is difficult to get a dial tone at times. This is because your line finders may be overloaded in the 200–300 series. If switching cost is included as part of your rental of an individual telephone station, you should ask telco to install morc switches so that you have adequate talking paths for your people.

On the other hand, if your telco charges you a monthly rental rate for each switch they install and also requires that you sign a five-year contract when you add switches, it would be better to try to rebalance your calling load.

The procedure is quite simple. Divide your people so that the light talkers and heavy talkers are evenly assigned to a dial level. In the example given, perhaps half of the purchasing and sales people would be given new three-digit numbers in the 400 series; a similar number of engineers would secure new numbers in the 200–300 group. While you will have to issue a new intraoffice telephone directory and perhaps pay a one-time wiring charge for the number changes, those costs will be far less than rental charges for additional switches. Further, many telcos do not charge for the wiring required to rebalance the dial system as they realize they may have erred in making the original number assignments.

## SECURING DETAILED TRAFFIC STUDIES

Telco has a responsibility to its customers to provide sufficient trunks and switches to ensure that their PABX system is operating satisfactorily. If you are receiving complaints concerning a slow dial tone, or if there are problems in completing calls once dialing commences, you should ask your telco to schedule a thorough traffic study of your system.

Sophisticated equipment is available to expedite telco's study. Electronic devices may be brought to your equipment room to evaluate switching equipment. Perhaps a photographic device will be used to record the position of all switches on a frequent scanning basis. Meters will be read from hour to hour or even minute to minute. Outdial trunk usage will be recorded.

When the observations and recording have been completed by telco's staff members, the data gathered are taken to telco's traffic engineers. They look at your existing trunks and switching apparatus and determine whether the traffic observation indicates the need for more apparatus or less. A recommendation is given to the telco marketing representative who serves you, and you then make your decision on the action to be taken.

As there is a trend in the industry to include the cost of power and switching in the monthly rental rate of the individual telephone station, it becomes telco's responsibility to provide adequate switching. If you are having problems under such a pricing arrangement, be sure to insist that the traffic study is made just as quickly as it can be scheduled. If telco is still charging you for individual switches, you must evaluate whether the indicated improvement in service to your users justifies your signing the five-year contract which will apply to any additional switches you may need. You must look ahead to your probable increase or decrease in telephone usage over the five-year period.

If telco recommends that you remove switches, be sure to ask yourself whether you will need those switches again when your business activity picks up on any seasonal pattern which applies. Once a switch is disconnected, a five-year contract may apply to its reinstallation.

## USING THE TELCO "OBSERVING SWITCHBOARD"

In some Bell System and independent companies, another free service is offered to help you appraise the service your switchboard operators and station users are giving to those who call you. This service is described in the section "Call-handling Observation" in Chapter 17.

## SELECTING A PROPER TRUNK CONFIGURATION

Many users of telephone service are not familiar with the types of central office trunks which are available. In some geographic areas, different types of trunks may have different costs. For instance, perhaps a trunk used only for incoming calls is priced at a lower rental rate than a two-way trunk which can handle both incoming and outgoing calls. Similarly, incoming and outgoing trunks may be offered only on a message-rate basis or only on a flat-rate basis, or perhaps you can mix both types of trunks to achieve the lowest cost.

### One-Way Incoming Trunks

These trunks appear on your switchboard. They can be used only by those calling you. Your operator cannot use them to dial outgoing calls. They are sometimes installed to ensure that talking paths are open for those calling you.

### Two-Way Trunks

These trunks appear on your switchboard. They are available for incoming calls, but if your operator wishes to dial an outside call, or perhaps award an outside line to a long-delayed WATS call, the trunks can be used in this way. If the two-way trunks are in sequence, the operator should always do her outdialing on the last numbers of the sequence. If nonsequential trunks appear on the switchboard, those trunks should be used for outdialed calls emanating from the switchboard.

### One-Way Outdial Trunks

These trunks do not appear on your switchboard. They are connected in your dial equipment room. The little aluminum packages of equipment are probably labeled "DCO-9" or some similar legend. Your equipment printout from telco will tell you how many trunks you have. Or you can count the aluminum boxes on the frames in the equipment room. Outdial trunks do not need sequential numbers. Any number will do. However, you cannot necessarily add trunks unless your dial system is "equipped" for the additional trunks. In other words, there must be space on the frames in the equipment room where the trunks can be mounted. If there is no space available, an order must be given to Western Electric for the needed equipment, which is usually furnished on long lead times. However, most telcos design spare mountings for equipment which may be needed in the future. It would be a good idea to ask your marketing representative to show you the spare mountings in which future trunks, line finders, selectors, and connectors can be installed.

### Two-Way Combination Trunks

These trunks appear on your switchboard. They differ from one-way incoming or two-way trunks because two separate lamps appear, one above the trunk jack and one below the jack. The brighter of the two lamps indicates that an incoming call is being received. The dimmer lamp does not blink, but remains steady. This lamp indicates that the trunk is being used for an outside call by someone who has dialed "9."

Thus, you can see that a two-way combination trunk works harder than the other trunks I have described. If all your incoming trunks are busy, and your two-way combination trunks are sequentially numbered, the next incoming call will flow into the two-way combination trunk group. Conversely, if all outdial trunks are busy on the ninth level, the next person dialing "9" will seize a two-way combination trunk and the lamp will glow at the switchboard. This indicates that the operator should not use that trunk if she should wish to give someone an outside line.

We recommend the use of two-way combination trunks to most users of dial service. The trunks work harder, so less trunks are needed. Telco's revenue is decreased, but many telco engineers indeed recommend that such trunks be installed. These trunks may or may not be separately identified on your equipment printout. The best method of determining if you have them is to look for the double-lamped trunk on your switchboard. Or you can go to your equipment room and see if you find any aluminum trunk boxes which are marked "2-way combo" or some similar wording.

If you do not now use such trunks, tell your telco marketing representative that you wish to decrease your trunk expense by installing them, disconnecting a greater number of two-way and one-way outdial trunks. You may have to read your meters carefully after installation to make sure that your "mix" is correct. You may also wish to order a blocked-call study on your incoming trunks which will include the new two-way combination trunks.

## SECURING FURTHER INFORMATION ON TELEPHONE TRAFFIC AND EQUIPMENT ENGINEERING

The traffic engineers who serve the telephone utilities and the interconnect vendors use techniques and statistical tables which have been developed and tested over a period of many years. If you would like to learn the details of traffic and equipment engineering, you may wish to acquire two books which contain descriptions of these techniques and the needed mathematical tables. The books, Roy Stone's *Cost Reductions in Wire Communications* and Gilbert R. Brackett's *Telephony's Telephone Traffic Engineering Handbook,* are listed in the bibliography.

## CORD SWITCHBOARDS

Tariff reading is a "must" if you are choosing a switchboard to operate the new dial system you are installing in your plant and office. You should also read the tariff if you need an additional switchboard position for your existing dial system.

Many different types of switchboards have been made over the years by Western Electric. One of the cord switchboards which was being manufactured in 1973 was the 608D (Figure 1). Other switchboards are sometimes manufactured or remanufactured for special situations where replacement of existing switchboards would cause hardship to telco's customer.

Consequently, you should be familiar with the phrase "manufacture-discontinued" which appears in tariffs and is also used in conversation by your telco marketing representative. It is only natural for the utility to sell you the "top of the line," and the

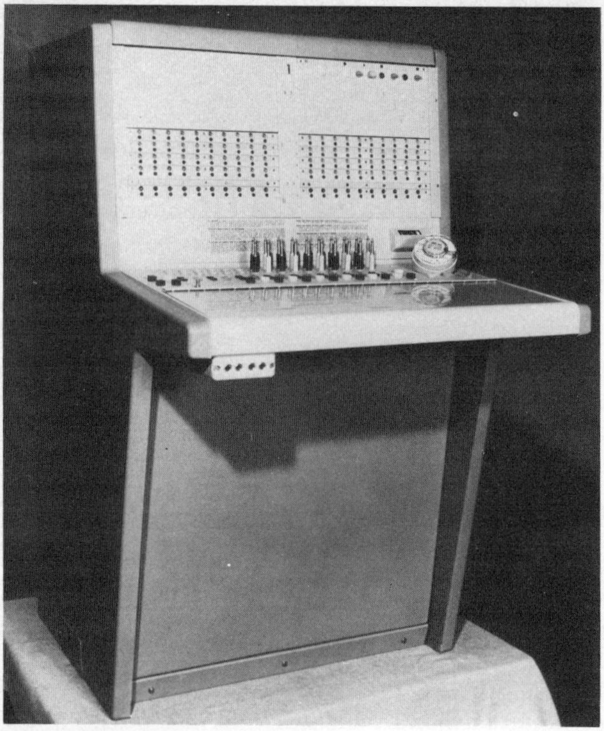

**Figure 1.** A 608D manual or dial cord switchboard. *(AT&T Co.)*

608D cord switchboard is an excellent piece of equipment. It has a modern appearance with a low profile and comes in a beige color. When the operator places the cord in a station jack, that telephone will ring automatically until it is answered. The 608D features automatic recall of the operator so she may transfer a call. The station user depresses his instrument's carriage button only once. A signal appears at the switchboard and the operator asks for your instructions. The 608D has a digital clock, which helps the operator time toll or WATS calls. It has a built-in PC mechanism, which helps her count calls.

The 608 models carry a substantial rental fee, which in 1973 was in the $125 to $175 range per month, whereas the older cord switchboards rent in the $50 area. In some Bell System companies, special packages are available which permit automatic ringing and recall mechanisms to be connected to the older-type switchboards. An extra charge is made monthly for the special features, but the added cost may be considerably less than the 608 rental fee.

So, *read your tariff.* While the tariff may say "manufacture-discontinued" where the older switchboard is described, the tariff may also have a footnote which says "available only for existing installations." This means that if you already possess one or more of the older boards, your telco is obligated by the tariff to furnish an additional board to you if you wish to expand your system.

Where will they procure the older unit? It may be sitting in their warehouse, where it was stored following their selling one of the 608 models to another customer. Quite possibly the stored unit was remanufactured or reconditioned by Western Electric before it was put in storage. All telcos have warehouses in which "Class C" (reconditioned) equipment is held for possible use in expanding the older systems. If your telco does not have the unit you need in their own stock, they have access to other Bell System warehouses around the country.

## CONSOLE SWITCHBOARDS

Most small telephone dial systems are operated with a console switchboard rather than a cord switchboard. The console resembles the Call Director, described in Chapters 7 and 11. If traffic requires a second operator, another console can be added. Modern "switch-loop" consoles work as a team. Incoming calls automatically are routed to an available console. In some installations in which older equipment has been used, incoming calls may not be routed uniformly to an available console position. In such cases, unanswered calls may be backed up behind one console while the other operator is idle. If this situation occurs, you should ask telco to revise their engineering and change your system.

In Chapter 12 on interconnected telephone systems, you will read about the features of consoles which operate the interconnected system. The Bell System has introduced a new console which is designed to meet competition. It features switch-loop operation and also mounts the Direct Station Selection (DSS) buttons in the console itself rather than providing a separate button panel. If you have a console-operated dial system and you believe you require a second or third console, it would be wise to ask your marketing representative for information on the newer equipment. At the same time, call the independent vendors and ask for details on their console-operated systems which interconnect with the utility's network.

Western Electric has developed a new console switchboard

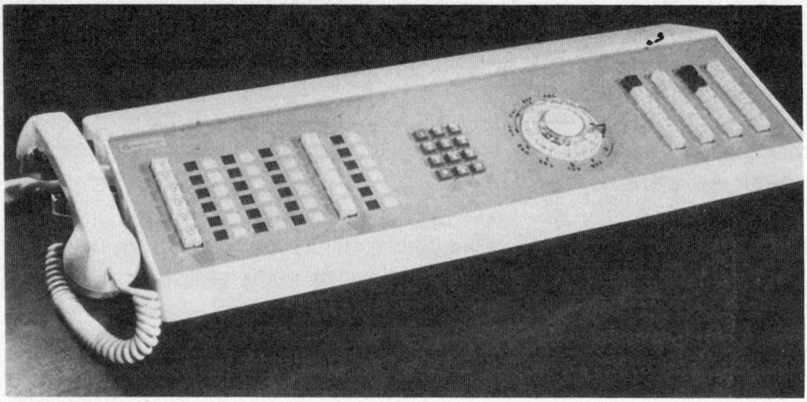

**Figure 2.** A console switchboard used for 812A PABX systems with a capacity of up to 2,000 lines. *(Western Electric Co.)*

(Figure 2) which is being used for operation of the 812A crossbar telephone system. This system has a capacity of up to 2,000 lines.

## SEASONAL EQUIPMENT DISCONNECTION

If you operate a business with requirements for telecommunications services which vary widely at different times of the year, you may wish to check telco's tariff concerning seasonal disconnection. For instance, a summer-only amusement park does not need its trunk, incoming WATS, and multiple switchboard installation during the winter season. In contrast, a department store needs its maximum number of facilities at Christmas, as well as at other times when special promotions are planned. Most telco tariffs provide for special rental rates during the off-season for such enterprises. Equipment is left in place, but rental stops or is reduced until reconnection takes place. In addition, the cost of reconnecting the facilities may be quoted on a wholesale basis rather than on a per-line or per-switchboard basis. However, if such special tariffs do not exist, it may not save much money to discontinue a facility for a few months because installation and reconnection charges have increased substantially in recent years.

## KEY PULSING AND TOUCH-TONE OPERATION

Many switchboards and consoles are equipped with rotary dials. The operators use these dials when they are placing outside calls. In some of the small package dial systems, the console operator must use the dial to connect an incoming call to the station user.

Other switchboards and consoles are equipped with a key-pulsing unit which sends a dial impulse to the equipment, but which is faster than a rotary dial. If you do not have Touch-Tone facilities in the central office which serves your dial system, ask telco if they can equip your switchboards or console with key-pulsing units.

The word "Touch-Tone" is a registered service mark of the Bell System. Interconnect manufacturers have their own trademarked designations for their particular system. A Touch-Tone telephone is a splendid method for dialing calls. In industrial engineering studies, we have found that a Touch-Tone "pad" saves 50 percent of dialing time for an inexperienced user. In most telcos, Touch-Tone calling is offered only if the central office will accept the Touch-Tone impulse. An extra charge is made for this service. In 1973 a typical charge was $50 monthly for the first 25 instruments equipped for Touch-Tone dialing and $1 monthly for each additional instrument. Touch-Tone calling can be a selective service. You can use both Touch-Tone and rotary dial service within the same system. Consequently, you may wish to analyze the work responsibilities of your people. Is a credit and collection clerk constantly on the phone? You can justify the extra $1 monthly cost in saved time. Similarly, a high-salaried executive who dials his own calls may justify the $12 yearly fee. Many organizations, recognizing the political problems of giving Touch-Tone service to one person and not to another, have converted all telephones to Touch-Tone operation. It is up to you to make the decision.

Touch-Tone pads can be mounted on switchboards. Also, when used in connection with the push-down card dialer (see Chapter 11), the Touch-Tone telephone can markedly increase work output in programmed calling. Such calling is frequently preplanned for order solicitation or credit and collection work.

The Touch-Tone telephone is an excellent device for keying numeric and alphanumeric data into a computer which may be accessed through the dial telephone system. The Touch-Tone pad has two extra keys which can be used for special data purposes.

Manufacturers of interconnected equipment offer key-operated dialing equipment on all systems as an extra-cost option. The advantages are similar, but the user must be cautious as there may be no speed advantage in the completion of outside calls unless telco's central office is equipped to accept the key-operated impulses. If the central office is not so equipped, special equipment slows down the impulse to rotary dial speed. On the other hand, the interconnected "speed calling"—as one manufacturer calls his optional feature—is indeed helpful for speeding internal calls.

## APPLICABILITY OF THESE IDEAS TO EQUIPMENT FURNISHED BY INDEPENDENT UTILITIES

This chapter has been devoted to suggestions which are applicable chiefly to localities served by the Bell System. If you are in an independent telco territory, ask your telco marketing representative to stop by and explain how his equipment differs from what I have described. In general, these ideas should still be helpful to you, as telephone equipment performs similarly regardless of its make.

## SOME SUGGESTIONS ON TELEPHONE COURTESY

The suggestions you have been reading concern mechanical matters for the most part. Your PABX system is run for the benefit of your telephone users, who are dependent on the patrons of your enterprise for their continuing employment. It is important that your people use good telephone manners if your telephone system and your enterprise are to function properly. Here are a few brief suggestions.

1. *Answer your own telephone.* Your caller wants to talk to *you.* If you are available and not talking with an outside visitor, don't permit your secretary to screen your call or even to answer it. Your callers may think that their time is more valuable than yours and may resent being asked questions by your secretary. If the call has been misdirected to you, courteously transfer the call yourself.

2. *Screen calls properly in a conference situation.* If you are in conference, ask your secretary to say something like this: "I'm so sorry, sir, but Mr. Brown is in conference now. May I have him call you when he is free? Or if your call is urgent, perhaps I can try to interrupt him." This shows that your secretary really wishes to help. She can always report in a few seconds that she could not interrupt you, and she can offer to find someone else to help.

3. *Always identify yourself immediately when placing a call.* A good secretary will not give information to an unknown voice. You will save much precious time if you always introduce yourself immediately to the person who answers the telephone. See the suggested phrases on p. 14 which help speed DDD calls to destination.

4. *Train your people in correct responses.* The phrases "He's not in yet," "He's not back from lunch yet," "He's gone for the day" are all common and reflect improperly on the executive's working habits. A simple response can be used: "I'm so sorry, sir, but he's away from his desk. I expect him back later this afternoon. May I have him call you, or may I find someone else who can help?"

5. *Don't permit people to wait indefinitely on a busy line.* Your secretary should offer to find someone else or arrange for a call-back.

# CHAPTER ELEVEN

# KEY SYSTEMS AND SUPPLEMENTARY TELEPHONE EQUIPMENT

Every user of telephone service can choose supplementary equipment to improve service and save executive time. The telephone utilities and the vendors of interconnected systems recognize the profit potential of such equipment and merchandise it actively. The alert telecommunications administrator will wish to study such equipment carefully to make sure that the presumed advantages justify the added cost. This chapter will help you to plan "key systems." This is the common term for groups of button-equipped telephones that permit one person to talk on another person's line. I will explain pricing trends within the telephone industry and describe the characteristics, advantages, and disadvantages of supplementary equipment offered in the tariffs of the telephone utilities. The independent vendors offer similar equipment. State and local taxes usually apply to the equipment described in this chapter. The federal excise tax does not apply to supplementary equipment associated with PABX systems.

## UNIT PRICING

For many years the Bell System and the independent telephone utilities have set a specific monthly rental rate for each item of telephone equipment. The total of the rates for all the units of equipment contained in an existing telephone system is given in the item "local service" found as a lump-sum billing on most monthly telephone statements. To analyze this monthly cost, the telephone subscriber must procure an itemized listing from his telco's business office. Such listings may be typed, merely showing the quantities of each item contained in the entire system with the total monthly billing for each category. More often, the information is given in a computer-produced form and shows the cost of each instrument and of all supplementary equipment. In some listings, the itemization of key system cost may be separate from the list of the numbered lines.

Under the unit pricing system, the customer may disconnect existing equipment or add additional equipment at will. As telephone service is billed in advance, a credit is due from the time the equipment is disconnected until that specific billing period ends. Thus, careful records must be kept of any verbal disconnect instructions given to telco. This helps you make sure that credits due you are shown on the separate statement covering "other charges and credits" that is enclosed with your telephone bill.

Equipment acquired from vendors of interconnected systems may be disconnected by the user or by a servicing organization, but such disconnection does not necessarily entitle the user to a credit on rental charges unless the contract covering the interconnected installation so specifies. Additional items of interconnected equipment can be procured easily from the vendor with an upward adjustment being made in rental, or an outright purchase of such equipment can be arranged.

## PACKAGE BILLING

When equipment is priced on a unit basis, the telephone utilities must do a great deal of detailed record-keeping. Each time a repairman changes the equipment configuration, corrections must be made to business office billing records. Over a period of time, errors occur. When the telephone user conducts the necessary periodic inventory, he will find items billed for which no equipment is present. When equipment is missing, the telephone utility should be required to refund the rental erroneously charged, making the refund retroactive to whatever point in time can be negotiated by the parties.

To cut paperwork and labor cost, telcos are now adopting package billing systems which set a fixed price for several features that may formerly have been priced separately. For instance, instead of charging for each individual push button on a six-button telephone instrument, telco may make a flat charge for a button-equipped telephone, regardless of the number of buttons which actually function on any particular instrument. Similarly, rather than charging for each individual neon lamp underneath the buttons, telco may make a flat charge monthly for an illuminated instrument. Instead of permitting the customer to specify the type of illumination to be used when a multibutton instrument is needed, telco may offer only the most expensive kind of illumination. Such a package includes a flashing lamp when the line is occupied and a rapidly flashing lamp when the line is placed on "hold." Many telco tariffs have a provision that if more than two working buttons are provided on any single telephone instrument, all buttons must be illuminated. Thus, flashing relays are provided for all the lines, adding substantially to the cost.

Because these pricing arrangements vary widely among the telcos, the administrator must learn how to read the public tariff found in the principal business office of his telco (see Chapter 6). Only in this document can you determine whether there are ways of escaping the requirement that you must rent equipment which you do not believe you need. For instance, you may decide that an executive and his secretary should not require illumination under the two buttons on each of their two telephones. Instead, audible signaling can be used by having their respective bells toned differently. In almost every package billing plan, such a nonilluminated package is offered. However, this alternative may not be described by the telco salesman because illumination brings his company much greater revenue.

Suppliers of interconnected equipment offer features and equipment very similar to that of the telephone utilities. The user must make similar decisions on utility of features contrasted with cost. In general, there is less "package billing" in the interconnect industry. Features can be ordered as needed without the user being forced to order something which is not required.

## EXTRA-COST EQUIPMENT

Now let me describe the extra-cost equipment which can be added to your basic telephone system.

### Holding Equipment

Because almost every button-equipped telephone has a "hold" button, you may assume that this is a standard feature of any multibutton telephone instrument. That is not true, as holding equipment is an optional, extra-cost accessory. Such equipment may be included as an integral part of a package system. If so, you may not be able to reduce cost by eliminating the holding feature. On the other hand, if your telco still uses unit pricing, you may find that there is a charge for the hold button itself and still another charge for the relay in the equipment which actually holds the call. For this reason the analyst should ask the telephone user if he requires a holding facility for a particular telephone line. While practice varies, normally a charge is made for holding on a per-line basis rather than on a per-instrument basis. Thus, you rent one relay per line, but the hold buttons themselves may be charged separately. Regardless of these details, you should determine whether a particular line requires a holding facility. If it does not, and if you are being charged on a unit pricing basis for holding equipment and hold buttons, you can reduce cost by discontinuing the holding feature.

### Pickup Buttons

Some telcos bill working buttons associated with a numbered telephone line at a flat monthly charge. A typical billing is 40 cents monthly or $4.80 yearly. If such a charge is made by your telco, you should inventory all working buttons and make sure that they are indeed needed by the user. If not, disconnect the button. There is never a charge for disconnecting any item of equipment furnished by a telephone utility. While a 40-cent push button may seem relatively insignificant in comparision to the total cost of a large telephone system, substantial cost reductions can come about from discontinuing the many unused buttons which have crept into the system over a period of time.

### Illumination

The telephone utilities receive very substantial revenue from their merchandising of illuminated circuits. There are different types of relays associated with such illumination. In general, if more than three lines must be monitored by a single individual, some type of illumination is required to identify which button should be depressed to answer an incoming call. Likewise, a noisy area without partitioned offices may require illumination so that a person can glance across the room to see if the ringing sound he has heard is actually his own telephone, which is easily identified by the flashing light. However, there are alternatives to this expensive illumination.

### Audible Signaling

Modern telephone instruments are equipped with a bell on which the tone can be adjusted. A knurled wheel on the bottom of the instrument casing permits you to vary the intensity of sound. The case of the telephone instrument likewise permits interior installation of a buzzer. Thus, at a secretarial position, it is possible to wire the telephone so that a bell signifies the primary line is ringing and a buzzer signifies the secondary line is ringing. Some telcos do not make any extra charge for an extra bell or buzzer located at a secondary position which reminds the secretary to answer the executive's telephone. If a charge is made, it is nominal. If a buzzer is required for an intercom circuit, then a separate bell can be installed near the secretary's desk. These supplementary bells are also controlled by a device which permits the sound to be regulated. The secretary can listen for a loud bell, a soft bell, or a buzzer and depress the appropriate button to answer calls. You can be sure that any secretary about to be deprived of her "lights" will say: "How will I know when my boss is talking on his line?" To counter such objections, read the sections in this chapter entitled "Exclusion Keys" and "Position-Busy Relay."

### Visual Signaling

Two basic units of equipment furnish visual signaling. One is a relay, which provides the facility for a flashing or steady light. The other component is the neon lamp itself, which usually appears underneath a push button on a key-equipped telephone. Lamps may also appear in "beehive" form, mounted on a desk or wall, or in a panel containing multiple lamps of different colors. There is a growing trend to include visual signaling as part of a package plan. Thus, if more than two lines appear in a specific group of related telephones, normally called a "key system," the customer is required to pay for illumination of all the lines in the system. However, if there are only two related lines, audible signaling may be substituted for visual signaling, with substantial savings.

Some telcos bill the customer for each separate neon lamp. A typical charge for such an item might be 40 cents monthly. Other telcos may charge a flat rate for all the lighted lamps on a six-button telephone. A typical charge per instrument would be 75 cents. The relay may be billed separately for each illuminated line. A typical charge might be $1.75 monthly. In package billing a typical charge of $3 monthly may include both the relay and the neon lamp. Or package plans may be priced according to the number of illuminated lines in the system. The analyst can secure details by inspecting the section of the tariff which governs key systems.

**Wink-Hold.** The standard type of illumination usually furnished by telco consists of a relay which flashes the line lamp on an incoming call and then keeps the lamp burning while the conversation is in progress. Some telcos include wink-hold as part of a standard package billing without additional cost. Other telcos make an additional charge, such as 50 cents monthly. The wink-hold relay puts a rapid flash on the line lamp when a call is placed on "hold." This feature is very helpful in a customer service department where the supervisor wants to check constantly on the number of delayed calls. But in a two-line system serving an executive and his secretary, the wink-hold relay may be an expensive substitute for secretarial alertness.

A number of alternative illumination arrangements may provide acceptable service at reduced costs. I shall now describe them.

**Ringing Current Illumination.** Suppose that your telephone is located in a noisy area. For a small rental charge like 35 cents monthly, your telco will mount a lamp near your desk which will flash as your telephone rings. The lamp is energized by the ringing current of your telephone system. Now you have flashing illumination without having to rent an expensive relay. If a secretary monitors several lines, the lamps may be mounted in panels. Different colors distinguish the different lines. This service is very useful when a person does not have a secretary and the telephone is answered by a receptionist or clerk. When the line button is depressed, the lamp goes out. Thus, this system will not indicate if a person is talking on a line. Also, no provision is made for wink-hold. However, the answering person may hold the line if her phone is equipped with a hold button and if the line answered is equipped with a hold relay.

**Position-Busy Relay.** In many offices, an executive may have a private business line which is terminated on his telephone. The line does not go through the office dial system or the switchboard. It ensures privacy of conversation. Because of security precautions, the line does not terminate on the secretary's instrument. If she receives a call for her boss, it is helpful to know whether he is talking on his private line. Also, if his regular switchboard line does not have illumination, it is helpful to know if he is talking on *any* of the lines available to him. The position-busy relay is connected to the switch hook of the telephone. When the user lifts the receiver, the relay energizes a neon lamp which is mounted on a desk or wall or is within the multibutton telephone instrument.

The lamp is "steady" instead of "flashing." This relay is less expensive than the flashing relay. The lamp is billed at a small monthly charge, such as 35 cents. This device is useful when the analyst is trying to reduce overall illumination costs.

### Exclusion Keys

You may have noticed that in modern telephones one of the carriage buttons is white. It may be shaped differently from its companion button on the other side of the handset cradle. When properly wired, this button excludes listeners who might pick up the line at other locations. If exclusion is specified at the time the telephone system is ordered, there usually is no installation charge. The feature can be added to an existing installation for a small one-time charge. In addition to providing privacy for the executive, the feature helps the secretary, too. For instance, her boss always dials his outgoing calls on the secretary's line, which serves as an alternate line on his six-button telephone. By pulling out the exclusion key, he prevents the secretary from trying to use her line while he is talking. This arrangement keeps his line free for his incoming calls, and the secretary delays placing her own calls until her own line is free. The key can only be wired to one line on a multibutton phone. Any telephone instrument can be modified to permit use of the exclusion technique. When the conversation is finished, the line is automatically restored to the secretary's use when her boss places his handset back on its cradle.

### Multibutton Telephones

In Chapter 7 I described the way in which these instruments can be used by small businesses.

The Call Director has been one of the greatest revenue producers for the Bell System. Comparable units are available from the independents. There are several less expensive methods of providing backup service for an unanswered telephone, which I will discuss first. Some common telephone button arrangements are reproduced in Figure 1.

**The Six-Button Telephone.** This has a capacity of six lines or five lines with a hold button.

**Two Six-Button Telephones.** This alternative is frequently overlooked by the analyst. Two instruments occupy about the same amount of desk space as one Call Director. Assuming that a hold button is needed on both phones, the remaining buttons provide up to 10 lines of backup capacity.

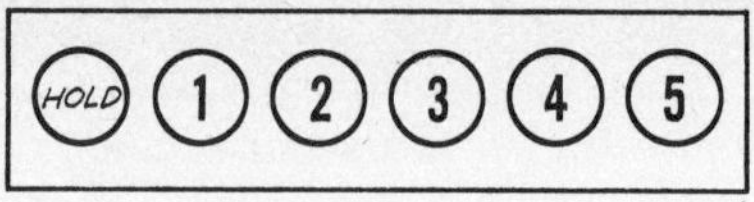


Requirement: Answer up to five lines.
Instrument: Six button telephone.

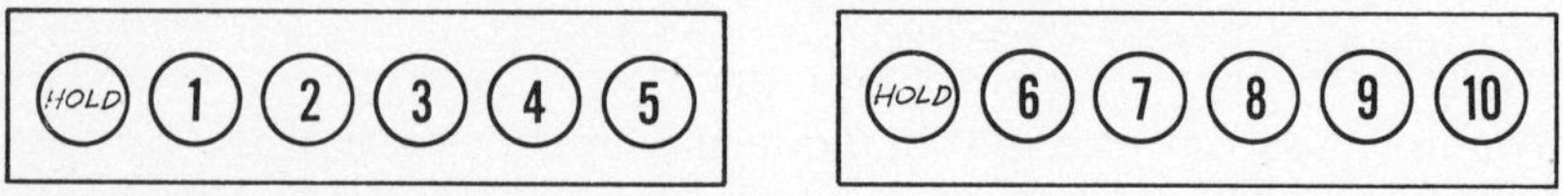


Requirement: Answer up to 10 lines.

Instruments: Two six button telephones which sit side-by-side. One of the two, or perhaps both instruments, may be "bridged" from lines which appear at other locations. Bridged instruments, sometimes called "extension" instruments, may rent for less than the *first* instrument associated with a numbered line.

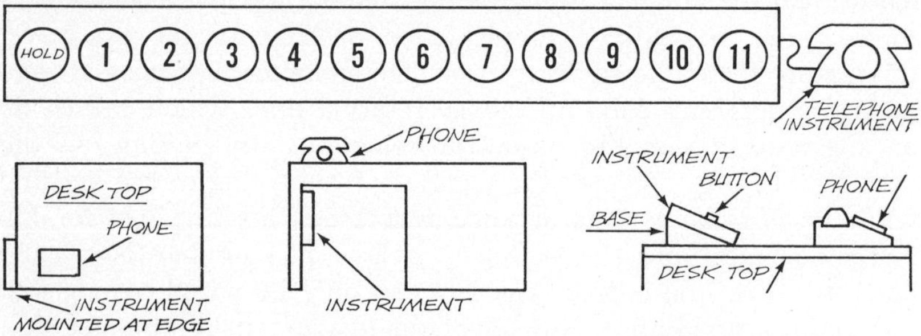


Requirement: Answer up to 11 lines.

Instrument: Twelve button Key Strip with associated telephone. May be mounted at edge of desk, in knee-space, or placed on a triangular base for desk-top use. This instrument is sometimes known as a Key Chief, or KU-2 or S-6051 Key.

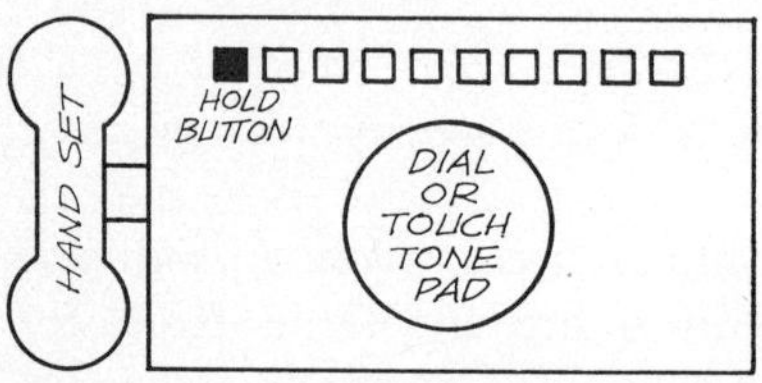


Alternate to the 12-button Key Strip:
Instrument: Ten button telephone. (Available only from certain Bell companies.)

**Figure 1.** Some common button arrangements. (The equipment illustrated is furnished and installed by Bell System companies.)

By choosing the multibutton telephone best suited to your needs, you may be able to cut down on the costs of these devices. Your final choice will probably be determined by cost, appearance of the units, and prestige requirements of the user.

In current tariffs, installation, move, and change charges have been increased substantially. Such charges are less for 6-, 10-, and 20-button telephones than for other multibutton instruments.

In the illustration, all telephones are equipped with hold buttons. This is an optional feature. If a telephone is used only for answering, and there is no need for holding a call on one line while talking on another line, the

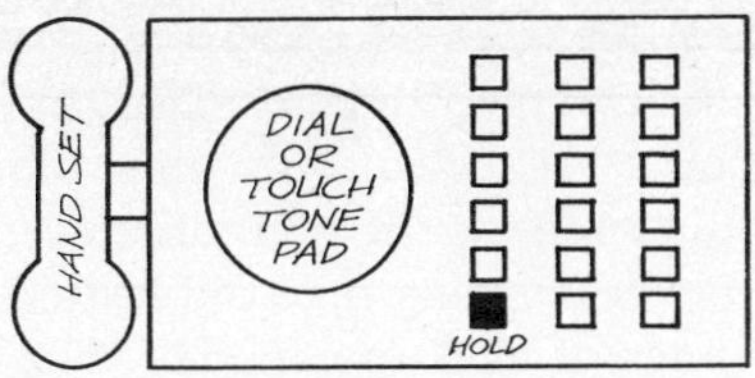


Requirement: Answer up to 17 lines.

Instrument: Eighteen button Call Director.

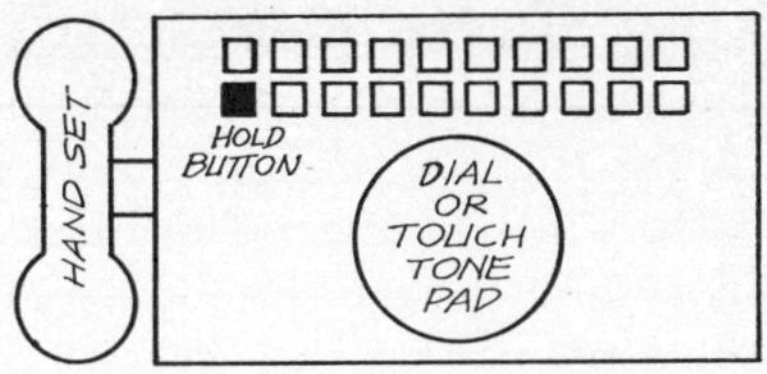


Requirement: Answer up to 19 lines.

Instrument: Twenty button telephone. (Available only from certain Bell companies.)

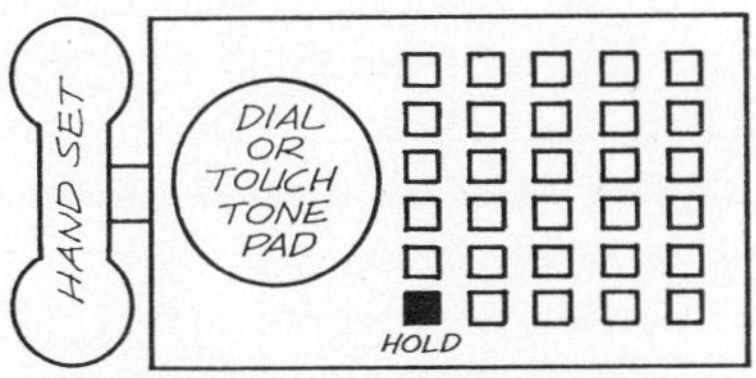


Requirement: Answer up to 29 lines.

Instrument: Thirty button Call Director.

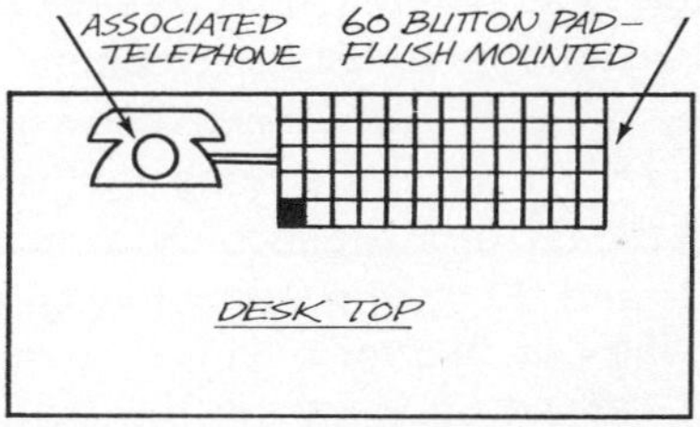


Requirement: Answer up to 59 lines.

Instrument: Sixty button Call Director. This instrument is usually flush-mounted in a specially built desk, or it may stand by itself in a large box.

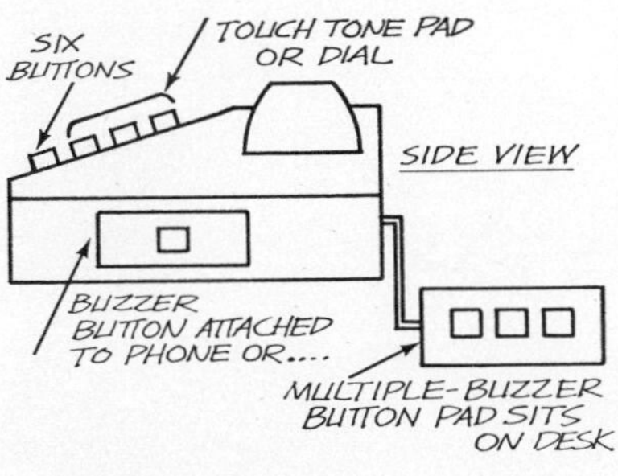


Requirement: Answer five lines, but also need one or more buzzer buttons.

Instrument: Six button telephone with buzzer pad located by its side. If only single button is required, attach buzzer pad to side of instrument.

hold button can be eliminated. This adds one line to the capacity of each instrument pictured.

Using two 6-button telephones may or may not be less expensive than using a 12-button key strip or a 10- or 20-button telephone, depending on the local tariff. To find out, read the tariff at your telco's business office.

If you are in a non-Bell territory, or if you are using an interconnected telephone system, the instruments available may be somewhat different. Certain utilities, for instance, furnish a nonilluminated three-button telephone; here, a separate hold button is provided for *each* line. To find out what your alternatives are, read your tariff or contact the marketing department of your interconnect vendor. *(Reprinted with permission from Administrative Management)*

**The 10-Button Telephone.** This instrument is a relatively new product of Western Electric. It is little larger than a six-button phone. Its installation charge may be a great deal less than that for a Call Director. This instrument is also being manufactured in a 20-button model (Figure 2). Thus, subtracting the hold button, these phones provide either 9 or 19 lines of backup capacity. They may not be available in some Bell System territories.

**The 12-Button Key Strip.** This device is a narrow rectangular box with a hold button and 11 line buttons. It is mounted on the side of a desk, in the kneehole, or on top. In some cities, a triangular base is available which tilts the keyboard toward the desk occupant. A regular nonbutton telephone is attached to the key strip. The key strip is sometimes known as a "KU-2," "S-6051," or "Key Chief." It normally rents for less than a Call Director. Depending on the tariff of the locality, it may likewise cost less than the newer 10- and 20-button instruments mentioned above.

**The Call Director.** This instrument comes in various sizes: 12, 18, 30, 60, or more buttons; an 18-button Call Director is shown in Figure 3. Sizes of 60 buttons or more may be placed in special mountings to bring the top of the instrument to a flush position

**Figure 2.** A 20-button desk telephone set. *(AT&T Co.)*

**Figure 3.** An 18-button Call Director with self-designating keys. *(AT&T Co.)*

on a custom-made desk. Rental charges are correspondingly greater as the size increases. Installation, move, and change charges are very high, usually running from a minimum of $25 to $150 or more per instrument.

**1A Key Equipment.** Prior to the development of the Call Director, telco furnished these boxes. Toggle switches mounted on top of the box handled one line in the forward position, another in the back position. Units were furnished to handle 6 or 12 lines, and many units could be mounted together for greater capacity. As these units are no longer manufactured, they are now available only as replacements or enlargements for existing systems, principally in the brokerage business.

**Choosing between Instruments.** All these units do exactly the same thing. To answer a call, the line button is depressed. If a hold button is installed, the line may be held while further information is procured on another line. Thus, choice between the instruments is determined by cost, appearance, and sometimes by the prestige requirements of the user. The analyst should read his local tariff to determine the rental cost and the installation, move, and change charges. For maximum economy, he should install the least expensive instrument. Historically, the Call Director has been recommended by telco's salesmen when the customer's requirements exceed the capacity of the six-button telephone. The thoughtful analyst can now consider the alternatives listed above. Instruments furnished by independent utilities and by interconnect vendors may be slightly different in capacity and appearance, but details can be secured easily by consulting the local tariff or by asking the interconnect vendor for marketing information and brochures.

### Hands-free Telephones

These instruments include a special loudspeaker which broadcasts the voice of the incoming caller. A microphone picks up the voice of the person called. The Western Electric model marketed by the Bell System provides excellent service for the larger conference room; independent utilities and interconnect vendors likewise furnish good equipment. However, rental cost is very high. In Ohio, for instance, rental cost in 1973 was $90 yearly. Consequently, all distributors of electronic equipment now sell competing equipment in the $15 to $25 range. Acoustic coupling is provided so that no wiring is necessary. The units are transistorized and battery-operated.

These speakerphones have acquired prestige status, yet at the same time they are genuinely useful for people who must write and talk at the same time. Also, they are helpful when one is waiting for a busy line to be free. Once the line is clear, the telephone handset can be reclaimed and conversation conducted normally. Most of these units distort voice quality and give a "barrel" sound. Because of their popularity, they tend to multiply rapidly within a company. I suggest that the analyst procure one of the purchasable units on a trial basis, demonstrate it to his executives, tabulate the savings of purchase versus rental, and then let management take its choice.

### Dial Intercoms

These little switching systems likewise produce large revenues for telco. One button on your telephone produces a dial tone on being depressed. When you dial one or two digits you are connected to someone else within your dial intercom system. Your call is announced by the sounding of a buzzer inside the telephone. The called person recognizes the sound as an intercom call, presses his intercom button, and talks to you.

**Types of Dial Intercoms.** These units come in models which permit either one conversation or two simultaneous conversations. Capacities range from a limit of nine stations on the basic model to greater coverage in two-path models. Some higher-cost packages provide for illumination so that people can tell if the line is being used. Another expensive package may provide exclusion so that no one can listen while a conversation is being held. In 1973, a typical yearly cost in Ohio for the least expensive nine-station switching mechanism was $51; the cost for the talking circuit was $18; and the cost for each buzzer in the circuit was $4.20.

**Purchased Equipment.** If there is a proven need for automatic intercom equipment, many manufacturers will be glad to quote on the purchase or lease of such systems. Either a separate telephone-type instrument or a voice-activated speaker system is required. The analyst should always weigh carefully the purchasing alternative when a complicated system is being considered.

For instance, an electronics supply catalog issued in midsummer 1973 offered a master station voice intercom with as many as four remote stations. For a nine-point circuit, two master stations would be required at the point where departmental calls are answered, and eight remote units would be located within the department. Exclusive of wiring and installation charges, the outright purchase of the equipment would cost $131. In contrast, the yearly rental of a nine-station dial intercom from Ohio Bell at 1973 rates would be $106.80.

Manufacturers of automatic intercom equipment continue to merchandise the "second telephone" concept. One telephone on your desk is used for outside calls; a second instrument is used for calls within your organization. You are able to reduce the number of switches rented from your utility, as the private system carries the traffic load of intraoffice conversation. Your cost for rental of equipment from the utility is decreased. Now that interconnected equipment is available, served by modern crossbar and electronic switching equipment, it will seldom be economical to operate two separate telephone systems within the same premises. Do your economic study carefully if you receive such a proposal!

**Uses of Dial Intercoms.** Here are some typical uses of this equipment:

*Executive override.* A department head may feel that his immediate need for an answer to a question is more important than the conversation which his subordinate is having on his regular telephone line. The buzzing of the intercom causes the subordinate to terminate or interrupt his conversation.

*Speeding dialing.* Some of these systems are sold on the basis of saving executive time: it is faster to dial one digit than three or four digits. This argument seldom makes sense when you compute the dialing time and contrast it to equipment cost.

*Miniature switchboards.* In some departments, telephones may be lightly used by departmental members. Possibly the "extension" or "bridged" telephone can be rented for less than it costs to supply a numbered line to each department member. Thus, perhaps four numbered lines serve nine people. A secretary answers all incoming lines. She places the call on hold, presses her dial intercom button, dials one digit, and instructs the answering person to pick up the incoming call on the appropriate line. The secretary thus becomes a subsidiary switchboard operator. If a Centrex system is involved, the cost of the dial intercom arrangement may be much less than renting a primary Centrex trunk for each member of the department.

## Manual Intercoms

Good telephone manners demand that the called person answer his own telephone without secretarial screening unless he is in conference. However, where screening is required, or where the executive is unwilling to dial his secretary through the company dial system, an executive-secretary intercom line is frequently installed. It terminates on one of the buttons of the phone. Another button activates a buzzer circuit. The answering secretary places an incoming call on "hold," presses the intercom button, presses the buzzer button, and announces the call. The executive presses the corresponding button on his phone to answer. Costs involve rental of the four buttons, the talking circuit, and the buzzer. In Ohio, the 1973 annual cost for such a two-person circuit was $32, including buzzer circuits in both directions. In contrast, electronics stores sell two-station voice intercom systems for less than $10, although such systems do not provide privacy for the talkers.

## Automatic Dialers

Many telephone users are impressed by new gadgetry which may seem to have executive time-saving implications. Dialers appear in many offices, but few users have done careful time studies on seconds saved. At the same time, dialers may be very useful in situations involving programmed calling for order solicitation and similar projects. Outside manufacturers are now marketing these devices for interconnection to telco's equipment.

**Card Dialers.** Plastic cards are provided; numbers are punched out. The card is filed in any order desired in a rack integral with the telephone instrument. The card is placed in a slot and pushed down; then the automatic mechanism takes over. This unit is fast, and it is particularly effective with Touch-Tone calling. The card for another call can be selected while the first call is being completed. Cost is less than for dialers featuring magnetic memory.

**Magnetic Dialers.** These units have various capacities of magnetic tape storage. They can be reprogrammed at will. The desired name is found and centered in a window; a button is depressed and the machine takes over. I suggest that the analyst contrast elapsed call-completion time in seconds with similar manual dialing time, as the older units are quite slow.

### Add-on Conference

This feature permits the telephone user to add on one or two additional people to a conversation without going through the switchboard operator. Special wiring is done on the buttons of his six-button instrument. While "add-on" is a standard feature on some of the newer dial equipment package systems, it is also available as an extra-cost accessory in the older dial systems.

### Switchboard Conference Circuits

Any switchboard and most switchboard consoles may be equipped for conference calling. Several types of circuits are used. The most popular "conference jack" permits an incoming caller to talk simultaneously to as many as four individuals within the dial system. If an outgoing caller dials the operator and asks to be connected to the conference circuit, up to three other people can talk simultaneously to the distant point. Cost varies from $6 to $12 monthly under current tariffs. Many of these circuits exist on switchboards and consoles, but executives may not be aware of their usefulness. The analyst can check to see if the circuit exists. Its usefulness can be publicized. If it is not used, it should be disconnected.

### No-Instrument Telephones

If you see such an instrument listed on your computer listing of the telephone equipment you are renting, you may assume that this is a numbered line which terminates on an instrument that is listed under a different number. Thus, the second line does not require an instrument of its own. Such lines are installed when an executive feels he must have two lines from the switchboard reserved for his exclusive use rather than sharing his second line with a secretary or some other person. If the sharing concept can be sold to your people, you may save a considerable sum by disconnecting the extra lines.

### No-Dial Telephones

Such phones may be used in areas where incoming calls are acceptable but no outgoing calls are permitted. In general, a telephone equipped with a dial is preferable. The dial telephone is "restricted" so that the user cannot dial an outside line. However, the telephone gives the user full ability to reach others within his dial system. In some states the no-dial telephone carries a reduced rental rate.

### Noise-suppressing Transmitters

These units replace the standard transmitter units contained in the mouthpiece portion of the telephone handset. They are directional and exclude noise from surrounding areas. Rental cost usually is around 50 cents monthly. They are very useful in computer rooms and in factory environments.

### Press-to-Listen Telephones

These instruments are also used in noisy areas. They cut off the transmitter when the button is pressed, clarifying incoming conversation by eliminating feedback through the phone. A word of caution: because of this feedback, both ends of a telephone may be microphonic. It is *not* sufficient to cover the transmitter with the palm of one's hand when talking in an "aside" during a conversation. Both transmitter and earpiece should be covered with *both* hands! This revelation may come as a shock to some of your telephone users—it is good advice to publicize!

### Telephones for the Hard of Hearing

These phones have an adjustment for volume of sound. Their monthly rental cost is minimal, but should the person with impaired hearing retire, it is well to disconnect the extra-cost unit. If you install one, keep checking on its use!

### Jack-equipped Telephones and Headsets

If a person's job constantly requires writing in the midst of conversation, a headset can be very helpful. New technology makes these units light and comfortable to wear. They plug into the jack-equipped telephone, which may cost $1 extra monthly for the jack. The headsets rent in the $3 to $4 range.

### Automatic Answering Equipment

Most telcos provide this equipment in several different models. After the switchboard closes, the main number can be connected to the automatic unit so that a message can be left for action the following day. Sophisticated equipment provides a portable device which an authorized person can carry with him. He calls the answering machine and holds the unit to the mouthpiece; then the machine plays back the accumulated messages. You may wish to compare the rental or purchase cost of such equipment with the services provided by a good "secretarial bureau," which provides you with human answering service for a monthly fee. Such service may be superior because the operator can refer a call to an executive's residence in emergencies. The makers of automatic answering equipment offer lease and purchase plans if you prefer to make a direct acquisition and interconnect such devices to telco's network.

### Call-transfer and Call-diverting Equipment

Many manufacturers advertise units which will hold an incoming call, permit a secretary to dial the executive's distant number on another line, and then "bridge" the incoming call to his residence. In early models it was hard to hear the conversation, but new miniaturized amplifiers are correcting the problem. In more sophisticated equipment, a diverting unit can automatically answer and hold an incoming call, dial another number, and connect the caller to the executive at the number he has programmed into the equipment. The diverter operates unattended. These units require interface devices which are rented from telco. The call-transfer and call-diverting units can be leased or purchased. They may be quite helpful for the small office which does not have a switchboard and wishes to avoid taking a message through an answering service or automatic answering machine, then having to make a call-back.

### Dictation Terminals

All major manufacturers of dictation equipment will be glad to show you how your telephone instrument can be used for dictation. By dialing a three- or four-digit number within your dial system you are connected to a telephone recorder. By single-digit dialing you can start, stop, play back your dictation, or indicate corrections. I have found that two telephone recorders may service up to 40 people, each of whom dictates only a few letters a day. The cost of the telephone relays and the recorders is far less than the capital outlay otherwisc required to acquire individual machines for each person.

### Access to Paging Systems

Most telcos are eliminating the rental of paging systems, preferring that these systems be furnished by others. However, any paging system can be connected into the telephone system. By pressing a hand switch or foot pedal the operator can speak into her regular headset or a separate microphone to page a person for whom a call is waiting. Also, the circuit can be arranged so that any person can dial a number which will connect him with the paging circuit, so he can page personally. The cost of the telephone connecting arrangement is minimal.

### Miniature Telephone Instruments

The Princess and Trimline telephones, which many people have in their own homes, are frequently found in reception areas and in the offices of executives. These instruments are billed at a rental rate which is higher than the rate for a standard telephone instrument. The analyst must point out such cost increments to management.

### Automatic Call-distributing Equipment

Airlines, department stores, and newspaper classified advertising departments frequently acquire this equipment from telco or from an independent manufacturer for interconnection to telco's network. Incoming calls are accepted in order, and if the call cannot be connected to a reservation clerk, a shopper, or an ad-taker, the customer hears a recorded message, either one-time or at intervals. Recorded music may likewise be used to soothe the waiting customer. The automatic call distributor is an expensive accessory, but it may be very helpful in maintaining good public relations.

### Miscellaneous Equipment

In the preceding sections I have covered those items of supplementary equipment that I feel are of most interest to telecommunications administrators. You can learn about other similar equipment, as well as its cost and availability, by examining the tariff filed in your telco business office. The tariff will give you many more ideas which may be very useful to your particular business.

CHAPTER TWELVE

# INTERCONNECTED TELEPHONE SYSTEMS AND APPARATUS

In 1968, the FCC announced its decision in the "Carterfone" case, ending the exclusive rights of telephone utilities to furnish telephone equipment to their customers. In 1969, AT&T issued tariffs to permit the interconnection of privately owned or leased equipment to its circuits. Since that time, eager manufacturers, distributors, and dealers have rushed into the marketplace to furnish telephone systems and peripheral equipment to users of telecommunications equipment.

As the market has evolved, the word "interconnect" has been applied to the entire range of services and equipment furnished by industry members. When you hear the word, keep in mind that interconnection is really not a new concept. In many localities, small, privately owned telephone utilities could not afford to purchase telephone equipment needed by large industries which happened to settle in the utility's marketing area. Through negotiation, the industry would purchase its equipment and interconnect it to the trunks furnished by the utility, thus relieving the utility of the necessity for raising capital to purchase the needed system. Under special circumstances, certain organizations have been allowed to purchase and interconnect their equipment in Bell System territories, too.

Also, it is important for you to know that some of the manufacturers now competing for your equipment needs have been competing actively for similar business for many decades. There is a strong and mature telephone equipment industry in the United States. Those manufacturers compete for the business of the independent telephone utilities. The equipment they manufacture has been tested and proven in the central offices of those utilities. Reliability is assured; trained servicing personnel roam the United States providing expertise to the utility customers of the manufacturers. It was only natural that these experienced manufacturers immediately saw the marketing potential of the FCC's historic Carterfone decision.

Manufacturers in other countries likewise recognized the possibilities of a new, lucrative market in the United States. Japanese, German, and Dutch makers of telecommunications equipment began to intensify their marketing activity in the United States. Companies abroad manufacture equipment of similar or even greater sophistication than that made by Western Electric and the suppliers to the independent telephone industry.

Thus, you, a consumer of telephone equipment, are the beneficiary of the most exciting development which has occurred in years of equipment technology research. There are important reasons why you should consider interconnection. There are equally important reasons why you must use extreme caution in selecting your vendor and in making your financial analysis. This chapter will give you a broad outline of interconnection, together with an explanation of its advantages and disadvantages.

## THE CARTERFONE DECISION

For years, the telephone utilities opposed the use of any equipment which could be attached physically to equipment, or used in conjunction with equipment, furnished on a rental basis by the utilities. Enterprising vendors sold devices which could help keep a conversation from being overheard. Shoulder rests for the handset were purchased widely. These devices were not electrical in nature, but the utilities tried to discourage their use. Other devices appeared which used an "acoustical" connection with the utility's equipment. Automatic telephone recorders picked up conversations by an inductive device which required no physical connection to the wiring of the telephone system. It was such a device which brought about the extensive hearings resulting in the Carterfone decision. The FCC broadened its decision to indicate that any device could be attached to the utility's telephone system, provided that the utility's equipment or network could not be damaged by any electric connections which were made.

AT&T then introduced a device called an interface to protect its equipment and network. The interface is a terminal point for connecting the common carrier's equipment to the customer-furnished equipment. In addition to protecting the common carrier electrically, it would seem that it is providing a sizable amount of revenue to the carriers. In 1973, in most utility territories, a separate interface was required for each trunk connected to any interconnected private telephone system or individual piece of equipment. A monthly rental charge is made for each interface. In 1973, those charges ranged from a few dollars a month to as much as $50 a month. In addition, the federal excise tax applied in 1973 to the interface devices required for connecting customer-owned equipment to the utility's facilities. It seems evident that some utilities priced the interface on the higher basis to discourage customers from patronizing the utility's new and quite aggressive competitors.

Consumer protests about the cost of interfacing developed a trend toward the "certification" of interconnected systems and devices, whereby the utilities would be required to set up the standards which must be met by the competing equipment. Once the private vendor met the standard, presumably his equipment could be connected to the network without the interface. A substantial saving in rental charges would result from this development. Advocates of this approach say that the common carriers are utilizing equipment manufactured by the same vendors who are marketing interconnected telephone systems. No interfaces are required between the utility's own equipment and that which the utility has purchased from the manufacturers. Thus, it does not seem right to require the utility's customers to rent the protective device. Most observers of the industry believe that the interface requirement will gradually disappear as certification procedures are established.

## BASIC REASONS FOR CONSIDERING INTERCONNECTION

The alert telecommunications administrator cannot afford to ignore the possibilities of service improvement and cost reduction which interconnection may bring about. Here are some of the important reasons behind this statement.

### Protection against Rate Increases

The FCC has jurisdiction over all rates for telephone service which apply to interstate telephone service. A tariff for interstate long distance or WATS service must be approved by the FCC before AT&T can put the rate into effect. In contrast, the amount paid by the consumer for the rental of his telephone instrument or system is set by the telephone utility serving his particular city. That rate is not subject to FCC jurisdiction. Instead, it is subject to regulation by a state regulatory body, normally identified as a public utility commission. In Texas, regulation of utility rates was managed by city governments in 1973.

Members of state public utility commissions are political appointees. A commissioner must necessarily be responsive to public opinion. As a user of residential telephone service is also a voter, the commissions historically have favored lower rates for the residential consumer and higher rates for the business organization, which does not vote. Consequently, rates for equipment rental rise regularly and at a faster pace than do rates for residential service.

The acquisition of an interconnected telephone system permits the user to budget for equipment rental cost on a fixed price basis over the term of his lease or on a purchase and depreciation basis. Costs are known and predictable. The user does not have to guess about the trend of rental rates in the future. However, to the extent that many users acquire interconnected equipment, the utility serving that locality may be forced to raise rates for trunking facilities to compensate for the loss of equipment rental revenue. Thus, consumers should watch for notifications of planned rate increases so that their voices may be heard by the regulating commission.

Another factor which may affect rates for telephone service is a possible change in rate-making philosophy. In the past, a utility established a uniform rate for furnishing telephone service that applied to the general geographic region in which its customers were located. Thus, a subscriber in a remote suburban location of sparse population paid the same rental rate as a subscriber in the densely populated downtown area. This is known as "average pricing." Critics of this system suggest that telephone rental service rates should be based on the actual cost of bringing the service to the point of use. If extensive cabling is required to reach a specific point, the cost of that cabling should be added to the rate charged that particular consumer.

This change is already demonstrated by substantial increases throughout the industry in the fees charged for installing, moving, or changing telephone instruments and systems. These costs have doubled and tripled in many areas. For instance, if a change is to be made in lines terminating on a multibutton instrument, the applicable "change charge" may be as much as $100 for the single adjustment.

The user of an interconnected telephone system ascertains the cost of rental in advance, and this amount will not change. Furthermore, the technology used by the interconnect manufacturers permits a quick connection, disconnection, or rearrangement of service through the use of plug-terminated instruments. This work can be done easily by members of the user's staff.

Another factor which may cause increases in the utility's equipment rental rates is the effect of the continual pressure applied by the FCC for a reduction of rates applicable to interstate communication. The utility may endeavor to offset revenue decreased by such lower rates through increasing rates for intrastate communication. In almost every state, intrastate long distance rates per airline mile are higher than interstate rates, sometimes substantially higher. The utilities have also tried to set rental rates which would return their investment in equipment within a three- to five-year period. In some cases, the customer is asked to sign a contract which ensures that a defined amount of rental will be paid over a three- to five-year period.

In contrast, the vendor of interconnected equipment may offer a fixed-price rental contract over a longer term. Ten-year contracts have been common. Financing is arranged through captive finance companies or from third-party lessors. As a result, the quoted monthly rental rate for interconnected equipment may be considerably less than the quotation of the utility. If purchase of the equipment is being considered, a discounted cash flow study may show that the use of accelerated depreciation may make purchase superior to monthly rental through the utility or rental via the leasing contract offered by the interconnect vendor.

From this discussion, you can understand why many organizations are examining acquisition of an interconnected system primarily to reduce the current level of expenditure for rental of their telephone equipment.

### Reducing the Cost of New Construction

Another important interconnect advantage is that of lowering the cost for new building construction. When a telephone utility is asked to furnish a system for a new building, it usually will require installation of a special conduit for its exclusive use. The telephone cable is placed in that conduit and must be installed separately from other electric cables. The cabling for an interconnected system can be placed in a conduit which carries other low-voltage wiring, such as that required for paging systems, music, radio, or television. The electrical contractor can "pull" all the low-voltage wiring at one time.

### Securing Better Service from the Utility

In past years, many organizations have had difficulty securing information and assistance from members of the local telco's staff. The consideration of an interconnected telephone system usually rectifies this situation rapidly. When studying interconnect, the administrator should notify his utility that a possible change is under consideration. Telco's marketing staff will react quickly. It is quite possible that service will improve so rapidly that no change in telephone system may be necessary. Or the utility may offer a new service which can be compared to the interconnect offerings.

### Adapting Telephone Systems to Changing Conditions

If your organization is dynamic and growing, you may eventually reach the point when the demand for telephone lines exceeds the capacity of your existing dial system and switchboard. If you cannot add additional lines because your system is not expansible, you must remove the old equipment and install something different. At this point you should certainly investigate the offerings of the interconnect vendors as well as those of your utility.

Even if your utility's system has room for additional switches, and even if you can provide space for another switchboard position, keep in mind that both switches and switchboards may be subject to a three- to five-year contract in your locality. If you sign such a contract, you may hesitate to change your system within the contractual period. This is because most contracts provide for the payment of "termination charges" if the system is discontinued prior to the end of the contract. Thus, the enlargement or replacement of an existing telephone system should be undertaken only after you have studied the possible replacement of your equipment with a modern interconnected system.

There is considerable mobility of departmental personnel in the larger companies. A new building may be erected for housing a substantial number of people in a research and development department. The administrator should consider an interconnected system for the new building, and he should also ask himself if an interconnected system would better serve the people remaining at the original location.

In considering this subject of changing conditions, you should be aware of the fact that it is better to be generous in your estimates of future growth. Telephone systems manufactured for the utilities and by the interconnect people have fixed capacities. They may or may not be expansible by the addition of modules. If you underestimate your future requirements of stations and trunks, you may have to remove your new system before you have completed your payment schedule. Salvage value may be far less than the amount you have invested.

## Cost of Space Rental

Telephone equipment rented from the utilities may require a large "equipment room" with controlled temperature, ventilation, and cleanliness. This is because the older "step-by-step" switching apparatus is usually housed in open frames rather than in cabinets. Further, if the enterprise is apt to expand, space planners must provide extra space for additional switching equipment to be added later. If the system is to be operated by traditional cord switchboards, both current and future switchboard housing requirements must be planned, including nearby lounge facilities for the operators' relief periods.

In major cities, this required space may be very costly. Space rental may run from $5 per square foot up to $15 or even higher. Consequently, the telecommunications administrator is under pressure from his building and facilities executives to conserve as much expensive space as possible. Almost all interconnected systems utilize compact console operation rather than switchboard operation. The switching equipment is of more recent design than step-by-step switches. Crossbar (Figure 1) or electronic switching equipment is furnished. Compact enclosed cabinets house the equipment

**Figure 1.** A crossbar switch for a PABX dial telephone system. *(North Electric Co.)*

in a fraction of the space formerly required (Figure 2). Thus, in a metropolitan high-rent district, the space-saving aspects of the interconnected system may be an important advantage. If you have been considering Centrex because you do not have space for the traditional step-by-step equipment, you will wish to evaluate the space requirements of interconnected telephone systems, which require little more space than that required by Centrex "CO" systems (see Chapter 9). Savings in space rental or occupancy are a vital component of your interconnection economic study.

In considering the factors above, remember that the purchase or lease of an interconnected system involves a contractual commitment which may extend for a much longer time than arrangements which can be made with your local utility. Management should look ahead and determine whether it is likely that the business will continue to occupy its present quarters for the duration of the telephone system lease. If the system is purchased or leased, management must determine, in advance, the proportion of the contract price which is involved in cabling. This cost would be nonrecoverable if the business should vacate its quarters, unless that cost were offset by a depreciation recapture. While switching equipment and instruments can be moved to a new location, installation costs are

**Figure 2.** An electronic switching system for use in an interconnected telephone system. *(Stromberg-Carlson Corp.)*

involved. These potential costs must be estimated if there is a chance that the business may be moving.

## Equipment Features

In addition to the possibilities of cost reduction, you may wish to consider interconnected equipment because it offers operating features which are superior to those offered by your utility. For many years areas of the United States served by independent telephone utilities enjoyed the use of service features which were not offered by the Bell System. This was because the manufacturers serving the independent utilities had progressed somewhat further in research and development. The same manufacturers who serve the independent utilities are now major factors in the interconnect market. While some manufacturers will only quote on direct installations involving the largest telephone systems, others will gladly furnish smaller and medium-sized systems. The consumer now has a choice between his utility's offering and that of several responsible organizations which deal directly with the consumer or through carefully chosen distributors and dealers.

Foreign manufacturers likewise market intensively through United States distributorships. Some of them are manufacturing telephone equipment within the country. In almost every case, such foreign equipment likewise offers features which were formerly not available in most territories served by the telephone utilities.

In Chapter 9, I have discussed the speed of consoles furnished by both domestic and foreign interconnect manufacturers. This equipment may be three times faster than modern switchboards or consoles of less recent design. Such speed can have a decided effect on the cost of switchboard labor; see Figure 3.

**Figure 3.** The console switchboard of an interconnected telephone system. *(Stromberg-Carlson Corp.)*

In a later section of this chapter, I will describe the particular features of the interconnected systems which have been popular with users. However, it is only fair to point out that the Bell System, in particular, is intensifying its effort to compete with the interconnect salesmen. Excellent new switching equipment is being introduced which includes the more popular features of interconnected systems. As I will be suggesting in more detail later in this chapter, you should always ask your utility's marketing representative to bid against your specifications once you have determined the features you desire.

## CAPACITIES OF INTERCONNECTED SYSTEMS

In the beginning of marketing effort on interconnected systems, the medium-sized business was picked as an initial target. Such a business might require 50 to 200 numbered telephone lines within its dial system. In more recent times, the interconnect manufacturers have been willing to quote on major systems which involve thousands of lines. Such systems resemble the central office telephone systems which are used by some of the smaller independent telephone utilities. Some of these utilities may serve only a few thousand users of telephone service. You can understand why technology that was developed to furnish dependable central office service for 2,000 subscribers can be adapted quite easily to the furnishing of dependable interconnected telephone service to 2,000 employees housed in a corporation headquarters building. Further, this technology is not new. It has been proved reliable in years of usage by the independent telephone utilities.

Can the small office take advantage of interconnect? In general, if a business requires less than 50 lines, the economic study of interconnect must be done very carefully. It is difficult for a small business to secure leasing service over a long period. Leases tend to be shorter, and payments each month will be larger. The small business should carefully evaluate the outright purchase of an interconnected system because of the substantial cash outlay which would be involved. A small business frequently occupies rental quarters. Could the premises' lease be canceled, or could the lease expire without a renewal provision? At the same time, a number of responsible manufacturers are manufacturing a type of key system interconnect telephone systems which permit the telephone user to answer a call from any incoming trunk. Also, small package dial systems are indeed being advertised and sold. If you are operating a small business, or if your sales office managers in distant cities are receiving marketing visits from interconnect manufacturers, you should at least evaluate the situation, as the key system market has changed a great deal. See the discussion in Chapter 7.

Regardless of the size of the system involved, the financial analysis and the wording of the contract are of critical importance to decision making.

## FINANCIAL AND ANALYTICAL FACTORS INVOLVED IN YOUR INTERCONNECT DECISION

In choosing between the rental of your telephone system from your utility or the lease or purchase of the equipment from your interconnect vendor, you must consider the effect of the financing arrangements on the cash flow requirements of your business.

### Discounted Cash Flow

In almost every organization of reasonable size, some member of the management staff has studied the principle of the time value of money. In brief, the principle states that dollars which can be paid out over a longer period of time are worth more today than dollars which must be paid out immediately or within a short period of time.

As an example, most telephone utilities quote an installation charge for a new telephone system. Also, most utilities require that rental be billed and paid in advance. The interconnect vendor, however, usually quotes a purchase price or leasing cost which includes an element for the cost of installation. The cost of installation, rather than being paid for in advance, is spread out over the life of the leasing contract. Cash outflow may be substantial over the period of the lease, but the user of the system does not have to disburse a large sum at the beginning of the contract. Some vendors may tailor the length of the lease contract to the actual amount which the user *wishes* to disburse monthly. The user may be offered a purchase option at the end of the basic lease term, or the lease may be renewed at a negotiated rate.

To make a proper analysis of the various cash outflows which are required by your alternative lease and purchase plans, it is best to discount the future payments to present worth. Financial analysts use a number of different techniques, such as "discounted cash flow" or "present worth of future revenue requirements." The Bell System has developed discounted cash flow studies to assist marketing personnel in meeting interconnect competition. The interconnect vendor may likewise offer to make studies for you.

As each organization is very anxious to have you adopt the system being offered, I feel that you should try to secure independent financial advice from your own staff experts or from the professional accountants and consultants who serve you.

### Choice between Lease and Purchase[1]

If you are considering outright purchase of your system, keep in mind that telephone equipment manufactured by suppliers to the independent telephone utilities is engineered to rigid quality standards. The equipment will operate satisfactorily for many years. The utilities insist on high equipment quality because the regulating utility commissions require that dependable service be given to the utility's customers. Standards of service are usually set forth in the formal documents which originally established the right of a particular utility to serve its customers on an exclusive basis.

Because of the longevity of telephone equipment, terms of depreciation applicable to it may be quite different from those applying to other equipment. Your accounting people will wish to check the depreciation rules applicable to the equipment you are considering. The longer period of capital recovery may influence you toward selecting a leasing arrangement which can be better tailored to your requirements.[2]

If your interconnect vendor offers only outright purchase terms, you should be able to secure a long-term leasing proposal from one of the organizations mentioned under the classification "Leasing Service" in the yellow pages of your telephone directory. Also, many banks offer leasing plans to their customers. If your vendor offers his own leasing plan, be sure to compare the offered terms with those available from commercial or banking leasing sources.

If you decide to lease, remember that the lessor will wish to recover his entire investment in the purchase of equipment for you. He will also add a factor to the financing rate for profit and overhead. Thus, your lease contract normally is noncancellable. You will be obligated to complete the payments even if you discontinue using the system by vacating your premises.

Another important factor in negotiating lease terms is to provide

[1] For a complete discussion of the advantages and disadvantages of equipment leasing see Harvey Greenfield and Frank K. Griesinger, *Sale-Leasebacks and Leasing in Real Estate and Equipment Transactions,* listed in the bibliography.

[2] In 1973, the IRS issued a new ruling which stipulated that switchgear and station equipment specified in FCC Uniform System of Accounts, Equipment Class 235 would be subject to depreciation lives varying from 8 to 12 years.

that the cost of insurance is paid by the lessor and that the payment of taxes is likewise his responsibility. This is because the IRS looks at leases carefully to determine if they are disguised contracts of sale. An owner normally insures his equipment. He likewise pays taxes assessed against it. If an examiner finds that your lease specifies that you must furnish insurance and pay all taxes, your long-term lease may be reclassified as a purchase. This would require adjustments on your past income tax accounting which could affect profits reported in a prior year.

Understandably, the lessor is anxious to avoid the extra overhead cost of supervising taxes and insurance applicable to leased equipment. If you find that you must provide insurance and pay taxes, be sure that you consider the impact of these costs on the economic evaluation you are making of interconnected equipment. State property taxes may be applicable to your equipment. Be sure that such taxes are paid, as unpaid taxes could result in liens being placed on your telephone system. If your lessor is paying taxes, check to make sure that the taxes are indeed paid. Some financing organizations, finding themselves short of capital, may be tempted to use tax reserves for operating purposes, leaving your equipment exposed to tax lien possibilities.

If the vendor is insuring your equipment, be sure to secure a copy of the insurance policy. Remember that it must be issued and effective before the date that you take possession of the interconnected system or apparatus.

## Vendor's Maintenance Capability

The vendor's maintenance capability is vital to the selection of your interconnected telephone system. Maintenance arrangements vary widely. Be sure you know exactly what your vendor proposes. For instance, in 1973 one major manufacturer of interconnected telephone systems was offering a 10-year lease with full maintenance included. In other words, the monthly lease payment included the cost of parts and service. The same manufacturer likewise offered a 10-year renewal of the lease on identical terms. Other manufacturers offered free maintenance labor and required parts within the first year, but specified a price for a separate maintenance contract for the second and succeeding years. Usually such contracts were subject to an escalation clause for possible increases in the cost of labor and materials. Almost all vendors offered maintenance on a per-call basis if desired.

However, the fact that maintenance service is available and that its cost is specified is not enough. You must know the *length of time* it will take the repairman to reach your place of business. If you have an emergency, how many minutes or hours will elapse before your system is back in operation? Usually contracts provide that the vendor will apply his best efforts to repair a major breakdown within hours. Minor changes may be completed in a longer time, perhaps two working days. The speed of emergency repairs is a crucial analytical point in considering interconnect and in negotiating with the prospective vendor. Be sure, also, that you know the hours and days of the week when service people will be available to you.

If your maintenance is to be supplied by an outside contractor whose organization is separate from that of the vendor, be sure to conduct a careful investigation of the servicing organization's financial responsibility and length of time in business. Such maintenance arrangements may be highly risky unless the contractor is of proven ability and unquestioned integrity. If your vendor is a packager and merchandiser of interconnect equipment manufactured by others, the same cautions apply.

To reduce the cost of outside maintenance on a special call basis, you may wish to contract for preventive maintenance service. Most vendors will offer to inspect equipment on a quarterly basis or, if the equipment is located in a dusty area, much more often.

Another analytical consideration is the possibility of your furnishing maintenance service with your own staff. If you are about to acquire a sizable interconnected system of 600 lines or more, you should consider asking the manufacturer of your equipment to train one or more of your employees in maintenance procedures. All major manufacturers furnish complimentary training to your employees throughout the life of your system. The training is usually done in special schools run in the plant and offices of the manufacturer.

Once your people are trained, routine maintenance can be done quickly and inexpensively. Modern switching equipment is furnished in module form; a defective module or circuit board may be quickly removed and replaced with a new or reconditioned part. Telephone instruments connect to conduits with plug-in connectors which eliminate tedious wiring. However, even if you decide to do your own maintenance work, be sure that your vendor has backup labor available for the inevitable emergencies or for complicated repair problems.

If the construction of your building makes moves and changes difficult, you may wish to reappraise the economics of the interconnected system. For instance, if the moving of a telephone requires using a circular saw on a concrete floor, the resulting labor cost can amount to a substantial sum if moves and changes are made frequently. You should determine if your telco provides move and change service at a reasonable flat rate. If so, that cost may be much less than the cost of labor and materials for moves made by your own staff. While there is a trend to charge out this work at a figure closer to actual cost, you should still carefully evaluate your utility's service arrangements if you have a "difficult" building.

### Vendor's Negotiations with Telco

Before awarding your contract, determine what responsibilities the vendor will be undertaking in negotiating with telco. A reputable vendor will be glad to assume responsibility for making arrangements for the interconnection of his system with telco's trunks. The vendor should also take the responsibility of negotiating the purchase of existing telephone cable if his system is to be installed in your present building.

Cable installed by a telephone utility is its property. It becomes part of telco's "plant," and if your system has been operating for many years, the cable installation probably is fully depreciated on the utility's books. However, telco will certainly place a value on the cable. It is possible that the utility will indicate its cable is not for sale. At the same time, telco will realize that if the cable is its property, you can require telco to remove it at telco's expense—and to do so on a schedule which is convenient to you. Cable removal is difficult and expensive. Your vendor should apply his negotiating skill to purchase existing cable at a reasonable cost.

The interconnect vendor is familiar with the utility's tariff in your locality. The vendor will know what interface devices, if any, are required. As such devices may be in short supply, a reputable vendor will see that the needed orders are placed in time.

### Choice between Offered Features

The telephone utilities are sometimes criticized for their intensive marketing of expensive peripheral equipment. Line illumination, speakerphones, automatic dialers, dial intercoms, and bell chimes can easily creep into a user's telephone system when executives are impressed by the prestige nature of such deluxe equipment (see Chapter 11).

Similarly, the vendors of interconnected telephone equipment will show you some impressive features which can be added to your system at extra cost. You may wish to examine the list of features found later in this chapter.

Be sure that you critically analyze the value to you of each suggested optional feature. For instance, is it really good business to permit an executive to interrupt the conversations of his staff by using an extra-cost "executive override" feature? Is the "compressed dialing" feature going to save sufficient executive time to offset its substantial cost? Is the automatic toll-ticketing device necessary to ensure control of long distance and message unit costs, or can manual recording procedures be installed?

Before entering into the final contractual negotiations, be sure to confer with your top executives, your interconnect salesman, and your company experts and consultants. Evaluate each feature against its cost.

Many organizations already have purchased or leased interconnected systems containing features which are far more sophisticated and expensive than were needed by the user. Your careful analysis will reduce the risk of incurring these substantial additional costs.

### Future Technological Obsolescence

Another factor to be considered is the possible future technological obsolescence of the equipment you are about to acquire. Telephone equipment is long-lived. If you lease your equipment from an independent vendor, your lease will almost always be a "pay-out" lease. This means that your lease payments will continue until your vendor has recovered his full cost of manufacture, plus his overhead and profit. Your lease will also include a substantial financing cost, particularly if the leasing service is furnished by an organization other than the manufacturer of the equipment.

Thus, ask yourself: "Will I be satisfied with the features offered by the interconnected system, realizing that I must use the system for the full period of the lease, or until my outright purchase money has been recaptured through depreciation?" Or would it be better to undertake a month-to-month lease of the utility's equipment, or perhaps acquire a system available from the utility under a short three-year contract? You must really be sure that the interconnected system being offered you is truly a technologically advanced system. Remember that the Strowger switch, introduced just before the beginning of the twentieth century, is a very versatile piece of telephone equipment. It is quite possible that such switching equipment is behind the advertised features of your new interconnected telephone system. Be sure that you secure unbiased, knowledgeable advice about the equipment inside the cabinet before you sign that interconnect contract, or look carefully yourself!

## Business Stability

The stability of your business is an important factor in your decision to consider interconnect. If you lease the system, you must be confident of your ability to meet the payments over the length of the lease. Equally important, you must be reasonably sure that the size of your business will not decrease markedly. Your interconnected system is usually constructed in modules. If your business grows, perhaps another module can be added to give you extra capacity. But if your business contracts, the system may be constructed in a way which will not permit you to negotiate with the vendor for return of a module for some kind of credit allowance. If such a return should be authorized, the credit you receive will be much less than the normal depreciated value of the equipment.

In addition to your concern about switching equipment, you must consider the possibility of a decrease in your need for telephone instruments. Suppose that your purchasing group moves to a different building which is served by a utility's system rather than an interconnected system. You will have surplus instruments on your hands, and those instruments have value. Most major interconnect vendors will permit the return of instruments at an agreed percentage of their original purchase price. If the instruments were leased, the monthly rental will be reduced by a price per returned instrument, which should be specified in your lease agreement.

If your candid appraisal indicates that the past history of your organization shows a considerable mobility of people and departments, it may be better to rent your telephone system from your utility. There is never a charge for disconnecting equipment. Rental is reduced immediately for the equipment returned, and a credit is issued promptly for the unused rental which you paid in advance. Further, if a new feature is offered by the utility, it may be possible to add it to your current system without your signing a new contract. Or, if your current system is under contract and a new, improved system becomes available, the tariff of your utility may provide that "termination charges" (see p. 158) will not apply on the old system when upgraded into a new system. Historically, an owner of any kind of machine tends to keep the machine running for as long as possible, just so the owner feels that he is getting his money's worth. This viewpoint keeps many an organization from securing the advantages of modern, technologically advanced equipment. Going back to an earlier principle: Make sure that your interconnected system is *really* better than the system available from your utility. Your utility's rental plan and relatively short contract periods make it easier to take advantage of new features which will be coming from the laboratories of the utilities in the years to come.

## Catastrophe Protection

As owner of your telephone equipment, the utility is responsible for replacing all defective parts, and the cost of repair service is included in your monthly rental charge. In the event of catastrophe, the utility has large corporate resources from which replacement equipment and emergency labor can be secured. If you are a prospective owner or lessor of interconnected equipment, talk with your vendor about his resources in case of a major failure. Are parts available within hours? Within days? Is there a chance that a major part might have to come from abroad? Is labor available to install a completely new set of switching apparatus if a flood destroys your system? Can the "switch" itself be procured, or is your installation specially engineered from nonstandard modules?

## Vendor Selection

My final analytical factor is the most crucial of all. It concerns the proper selection of your interconnect manufacturer and any distributor or dealer organization which may be assigned by the manufacturer to service your installation.

Even in the comparatively short life of the interconnect companies there has been high mortality. Telephone systems are very expensive. A majority of customers prefer to lease equipment rather than to make a cash outlay which may run to hundreds of thousands of dollars. In turn, this places a difficult burden on the interconnect manufacturer to provide financing for himself and for his dealer-distributor organization. If his sales department is aggressive, orders may be received which tax the resources of the financial vice president. The manufacturing organization must either find additional financing sources or offer itself for sale to a larger company. If the latter alternative is selected, the company being acquired finds itself with new corporate management which may not view its acquiree's policies with complete approval. Management changes may be ordered, and the new management may elect to guide the subsidiary company or division into a different line of activity. In some major companies, senior management is insistent that each division produce a specified rate of return on the capital assets invested. If this goal is not met, the division may be sold or its direction may be changed.

All these risks may affect you as the consumer of the telephone services provided by the interconnect vendor. If the vendor is no longer in business, it may be very difficult to get repair parts, and even more difficult to find properly skilled persons to install them. The interconnect industry has expanded so rapidly that maintenance people are scarce. There is great competition for such labor. The repair department employees of an interconnect manufacturer will quickly relocate themselves if their employer goes out of business. If your equipment needs servicing, you may have difficulty in finding qualified people to do the job.

These problems can be lessened if you will do a careful job of investigating the experience and financial condition of the organizations which want to do business with you. There are many fine companies which have been making excellent telephone equipment for years. They supply the independent utilities, and they can be expected to remain in business for years to come. There are likewise splendid foreign companies which are even larger than the United States manufacturers. If you acquire either foreign or domestic equipment, be sure that adequate parts stocks are maintained in this country. And carefully check the caliber and experience of the organization which will maintain your equipment. Will the manufacturer send factory personnel directly to you within a matter of hours if his distributor fails or is unable to supply service in a reasonable time? Many organizations view interconnect as an opportunity to gain access to a potentially profitable market. Such companies may have little or no experience in manufacturing telephone switching apparatus. Be very careful when you consider doing business with such firms.

## STEPS TO TAKE IN CONSIDERING INTERCONNECT

In the preceding sections, I have described the major factors which must be considered before the decision is made to examine the possibility of acquiring an interconnected telephone system. Now I will present some specific instructions for your study.

### Inventory Present Equipment

In many organizations, no effort is made to verify if the amount on the telephone bill under "local service" is correct. That is the usual title of the charge for rental of your dial system. You should ask your telco for a copy of their computer listing which itemizes every item of equipment for which you are being charged a monthly rental. With some study and with assistance from your utility's marketing representative or your consultant, you can identify each item and then inventory your system with your own personnel. Missing items can be reported to the utility so that a retroactive credit can be prepared. The utility will start billing you for items of equipment which are present but are not being charged to you. When the inventory is complete, you will know the correct cost of the present system. Equally important, you will know what equipment you have and how your people use that equipment.

### Challenge Equipment Usage

In Chapter 11 you will find many suggestions for reducing the cost of key systems and peripheral equipment which raise the amount of your monthly bill. I suggest that you appoint an equipment coordinator and give him responsibility for a complete reexamination of departmental key system requirements. Try to eliminate unused buttons and equipment. Substitute less expensive equipment which may perform approximately the same function. When this work is done, your rental bill will have been reduced by a substantial amount. At this point you will have a better basis for making your economic study of an interconnected system. This is because you will have reduced the cost of your current system to essential cost, not deluxe cost.

### Determine What You Want

In a following section of this chapter, I will describe some of the more important features offered by the interconnect manufacturers. Any interconnect marketing representative will furnish you with a complete list of the features provided by his equipment. Management should appraise these features and decide whether they are worth their stated cost.

### Ask Your Utility to Modify Your System

As part of your interconnect study, call your utility marketing representative and ask him to visit you. Explain what you are doing. Describe the features which you feel would be helpful to your executives. Find out whether similar features are available within your local tariff, and, if so, whether they could be incorporated into your *present* telephone system. If not, encourage your utility representative to consult with his superiors in the marketing department to determine whether your utility is about to offer a system which incorporates such features. If it is not, ask him to determine whether his company is willing to contact manufacturers of the equipment you need so that the manufacturers may quote the utility on cost and availability. The utility can file a special tariff request with the public utility commission to cover the equipment being ordered for you. In this way you secure the needed features, yet you can continue to use the proven maintenance service of your utility. Further, in most states such special tariffs are not subject to the periodic general rate increases which apply to the utility's general line of telephone equipment.

### Draw Up Specifications

To ensure that you will be receiving quotations from interconnect vendors on equipment which will provide the exact service you require, you must draw up a specification in telephone language. This work requires the assistance of an experienced employee with telecommunications background or a qualified management consultant. Specification drawing is not a job for amateurs, as the specifications will determine whether or not your system will perform as you expect it to perform. You may wish to acquire a set of the specifications used by the government's Rural Electrification Administration when telephone equipment is being purchased. These specifications contain language which is understood by all major manufacturers of telephone equipment. If you do not understand the language in the specification, be sure to have a properly qualified person explain the meaning of each term. Then you can correctly adapt the language to the quantities and types of equipment you wish to acquire.

Here are a few suggestions on preparing the specifications.

**Type of Equipment.** I have explained the differences between types of telephone switching equipment in Chapter 8. You should specify the type of equipment you want: step-by-step, crossbar, or ESS. The cost of maintaining step-by-step equipment is greater, and it furnishes less modern features. Maintenance cost is about the same on crossbar and ESS equipment, but ESS has an advantage in a "hostile" environment, such as a dusty area.

**Station Equipment.** Be sure to specify the number of instruments required, segregated by type of instrument. If no illumination relays are required, the button-equipped telephones usually are available in 3-, 6-, 10-, and 20-line instruments. If illumination relays are required, remember that relay and nonrelay equipment usually cannot be mixed in a single key system. Also, you may find that wink-hold relays may not be available from your interconnect vendor. Remember that the sale or lease of telephone instruments and relays adds substantially to your vendor's income and profit. The cost-cutting techniques described in Chapter 11 will help you in drafting your specification.

**Switching Equipment Features.** Each manufacturer has his own list of the features supplied by his equipment. The list of features included in this chapter may help you interpret what your vendor offers or what you would like to have him offer. Describe what the feature should do for you. Be cautious about using the name of the feature without describing its function. This will avoid your using a specific manufacturer's trade name for a feature; that name might not be understood by another bidder.

**Trunks.** Be sure to utilize your crystal ball when you estimate your trunking requirements. Interconnected telephone equipment is long-lived. At the same time, it has defined limitations on the number of trunks and stations which can be accommodated by a particular machine. You should try to estimate your requirements for the next 10 years, as the manufacturer may offer you a 10-year lease. If you purchase, your equipment will function satisfactorily for at least 10 years. If you do not estimate your requirements correctly, and if your equipment's capacity cannot be increased by adding modules, you may have to remove the equipment before the end of its useful life or before you have finished paying for it. It is not necessary to *pay* your local utility for trunks before you actually install them, but your equipment must have the frame capacity to permit installation of additional trunks as you need them. Be sure to include your trunk requirements for WATS, FX, and tie lines.

**Maintenance.** If you have decided on a particular type of maintenance service which seems best for you, mention that type of service in your specification.

**Schedule of Work to Be Performed.** The specification should require the vendor to furnish a schedule showing the dates on which each phase of the contract will start and the dates on which the work will be completed.

**Send Specifications Out for Bid.** You will have determined which interconnect vendors are reliable and which can furnish an adequate grade of service in your vicinity. Send a set of specifications to each organization. Also, be sure to send a set to your local telco, as that organization may indeed wish to compete vigorously with its competitors.

### Analysis of Bids Received

If you have written your requests for bids in "telephone language" the bidders should quote on approximately the same equipment, even though different vendors have different names for typical components. For instance, the Bell System's registered service mark Touch-Tone may be shown in your bids as Tone-Dial or some similar name. Thus, you must carefully compare the bids with one another. Here are a few specific suggestions; others are found in this chapter's section on contractual arrangements.

**Check against Specification.** Has anything been left out? Remember that this is a highly competitive business. Your vendor may have decided that his bid would look better if he omitted something you have specified, taking his chances on convincing you that the item really isn't necessary or that you won't notice the omission.

**Check on Options versus Standard Equipment.** The bidders should be careful to specify which features are included as standard equipment and which are available at an additional cost. If you find that standard features are included which you do not feel are needed, ask your vendor if his price can be reduced by eliminating those features.

**Use Your Accounting Department to Analyze and Equate Quoted Cost.** There is no substitute for careful analytical technique when you are comparing bids on equipment which will require such a large cash outlay in the years to come.

**Check References.** Talk to a responsible *officer* in an organization which uses a system comparable to that you are considering. If possible, visit that office and physically inspect the operation of the apparatus. Be sure to talk with the switchboard operators as well as the administrative services manager. Also, verify the promptness and caliber of the maintenance service rendered.

### Prepare Documentation

As you would expect, most standard contracts covering an interconnect installation safeguard the manufacturer rather than the buyer or lessor. It is so important to prepare your contract properly that I am including a special section in this chapter which will give you some specific suggestions.

### Establish Liaison with Chosen Bidder

As manufacture of the system progresses, one of your people should work closely with the vendor's engineers, for questions will arise which deserve careful attention by both organizations. You should keep close watch on the progress of the installation. Check to make sure that telco will have the needed trunks available for connection at the time your system is ready. Make sure that your vendor has arranged for the needed interfaces, if required. If delivery of the interfaces is to be delayed, urge your vendor to negotiate with telco for connection of the trunks without interfaces. Your liaison representative should likewise constantly check the installation time schedule to make sure that the system will be ready on schedule.

### Test Installation before Payment

Your expert employee or consultant should carefully inspect all items delivered under the contract to make sure that every system component has been furnished. Your advisers and the vendor should work out testing routines to be utilized before the system begins operation. Do not permit your accounting department to pay any invoice from your vendor until you are satisfied that the system is working perfectly and that all components are in place and functioning as ordered.

## CONTRACTUAL ARRANGEMENTS FOR THE INTERCONNECT INSTALLATION

In this section, I suggest protective ideas which may benefit you if they are inserted in your contractual documents. You may be tempted to include some of these ideas in your request for a bid. I would advise you to confine your bid specification to the technical details of your system. Your vendor will then quote on the cost of the system as specified. The negotiation on wording of your purchase or lease contract comes later. It is better to let the vendor exhibit his standard contract before you start to specify your own documentation requirements. This is because a premature listing of your requirements may result in the vendor increasing his contract price to cover your special demands.

The major interconnect vendors have developed detailed standard contracts. They are used in purchase and lease negotiations, and samples may be furnished to you at the time the bids are mailed. A number of vendors use contracts based on the "Universal Commercial Code" (UCC). The UCC incorporates legal language which has been tested in the courts. However, that language is designed to benefit the *vendor,* not the *user.*

Consequently, it is important that you tell your vendor what *you* want included in your contract. As the vendor has already quoted you a fixed price for purchase or lease of the system you have specified, negotiation of a proper contract is now easier, because the vendor knows there is competition and he wants your business. Be sure to ask your legal advisers to assist you in drafting the contract and to participate in the negotiations. Your total commitment for your interconnected system is a very large sum of money. You must protect that commitment by securing the best possible advice from experts.

### Financial Arrangements

I have discussed the various advantages of leasing and purchasing in an earlier section of this chapter. Be sure that your contract lists any required down payment. If you are leasing your system, try to avoid any provision which requires that payment be made in advance for a certain number of lease installments. Your contract should provide that no payment will be made until you accept the system as being installed and as operating according to the provisions of the contract.

### Warrantees

Be sure that you specify how long the manufacturer guarantees the operation of the equipment. If the equipment fails, does the warranty cover only the replacement of parts, or does it also include the labor needed to install the parts? If you are leasing, does your rental payment include all needed maintenance labor and parts, or is there a separate maintenance arrangement at extra cost?

### Liability for Nonperformance

Be sure to require specification of the date on which installation will be completed. Penalties should apply if installation is delayed. Likewise, penalties should apply if maintenance service is not furnished within the agreed time limits.

### Assignments of Contract

If you are leasing the equipment or purchasing on a contract of conditional sale, your vendor may refinance your contract with a bank or insurance company. You should determine where *title* to your equipment actually rests. If a bank or finance company holds title, what arrangements have been made for that organization's assumption of warranty and maintenance agreements? Try to negotiate a contract provision which provides that you must approve any assignment of your contract.

### Successors

If your vendor sells his business to another organization, what happens to the agreements for warranty and maintenance? Your contract should stipulate that any successor organization must assume the responsibilities of the original vendor.

### Notices

The contract should specifically state the type of notice to be sent to you in the event of assignment or succession, and it should state where that notice will be sent.

### Detailed Description of Equipment to Be Supplied

The contract should list every item of equipment so that you can verify the equipment has been delivered before you approve payment.

### Maintenance of Equipment

I have given you a number of suggestions about maintenance in the preceding section on analytical factors in the interconnect decision. Be sure that the contract specifically covers the type of maintenance to be supplied, the cost of the service, and the hours and days of the week when the service will be available. You should insist on penalties being assessed against the vendor if response times in the contract are not met. The contract should cover your right to send your employees to the factory of the vendor for complimentary training in maintenance procedures. You should have the right to send reasonable numbers of employees to the school during the entire life of your equipment installation.

### Negotiations with Telco

Your contract should provide that all negotiations with telco will be conducted by the vendor, particularly negotiations for acquisition of existing cable and for needed trunks and interface devices.

### Cancellation of Service by Telco

Your contract should likewise provide that if telco attempts to disconnect your interconnected apparatus, refusing to render service, and if this action should be sustained by the public utility commission in your state, your vendor will buy back your system at its then depreciated value. In my judgment, the principle of interconnection will be with us for many years and will be sustained by the courts if challenged. However, the state commissions are jealous of their right to regulate telecommunications activity within the state. In at least one instance, a state commission endeavored to prohibit the use of interconnected telephone systems for communication on an intrastate basis. As any system uses central office trunks which can access both intrastate and interstate telephone numbers, it would not be feasible to install a telco system for intrastate calling and an interconnected system for interstate calling. A proper contract clause should help protect you from protracted litigation should this unlikely situation occur.

### Working Drawings

The contract should state that all working drawings such as floor layouts, cabling plans, and electrical blueprints will be your property. Thus, you will have vital information with which to direct your own maintenance, move, and change work. These prints should likewise be available to your personnel during the installation process, so your engineers and consultants can check on the work of the vendor's installation crew.

### Insurance and Taxes

For the several reasons explained earlier in this chapter, try to provide in your contract for the vendor to take the responsibility for paying the taxes and procuring proper insurance coverage. The contract should stipulate that the vendor will furnish satisfactory evidence of tax payments and also give the user a copy of the applicable insurance policy.

### Continuing Advice and Assistance

As you experience turnover of personnel, you will need additional assistance from your vendor in training switchboard operators and in conducting training meetings for new employees. The vendor's technical representatives should be available on a complimentary basis to discuss problems which arise, particularly those which may involve use of the interconnected facilities with the trunks of the utility. Your contract should spell out the vendor's continuing responsibilities for advice, assistance, and training.

### Adherence to Local Codes

Because of the variety of local requirements on occupational health, safety, and fire protection, your contract should provide that the vendor be familiar with local codes and that the installation be in full conformance with those codes. Correction of installation errors must be performed at no cost to you.

### Schedule of Additions and Deletions

What happens if your business expands markedly before your installation is complete, requiring more switching apparatus or instruments? Conversely, what arrangements can be made to return apparatus if you need fewer connected lines because of business changes? Your contract should provide a schedule of costs and credits for additions and deletions to the system made over the life of the contract.

### General Installation and Testing Provisions

If one of your staff members or consultants is an expert in specifying proper testing procedures for the system, make sure those procedures are listed in the contract. Contracts should provide that the vendor use workmen who are able to conduct themselves on your premises in a satisfactory way. The premises themselves should be kept clean and safe. Debris should be gathered and hauled away by the vendor, unless the cost of such service will add substantially to your contract price.

### Patent Infringement

While basic telephone equipment has been manufactured long enough to establish the proprietary rights of the various vendors, a patent infringement might somehow involve a user of an interconnected system as well as its maker. Your contract should contain one of the usual clauses protecting you against such an action. Your legal department can assist you in framing the proper language, just as they can assist you in negotiating the other protective clauses mentioned in this chapter.

## FEATURES OF INTERCONNECTED TELEPHONE SYSTEMS

Every manufacturer of interconnected telephone equipment is proud of the features that can be offered to you, the consumer. Some features are "standard" and are included in the system's quoted price. Other features are optional and are available at an added cost.

Your executives will be enthusiastic about acquiring many of the 78 features you will find in the list which follows. Be sure that you consider whether any added cost is justified by the convenience or proven time-saving of the feature. Interconnect manufacturers are in business to make a profit. Their marketing representatives are well trained. Give careful thought to the selection of your alternatives.

In general, a given feature must be selected before your system is manufactured. Once your switching equipment is put together, it may not be possible to take it apart to add some feature which appeals to a particular executive at a later time.

Before listing the features, it may be helpful to explain the meaning of terms you will be reading in the descriptions.

*Attendant.* The person who operates the console switchboard.

*Carriage button.* Every desk-type telephone has a switch which disconnects the conversation when the handset is returned to the main frame of the instrument. In most instruments, the handset rests on two carriage buttons. Either button, when depressed, breaks the existing circuit and restores the dial tone.

*Common control equipment.* The switching equipment which is activated by the dialing or pulsing of a call.

*Console.* The multibutton instrument which serves as a switchboard in interconnected telephone systems.

*Indication.* A visual or an audible signal.

*Off hook.* The telephone handset is separated from the body of the instrument, so that a dial tone can be heard, and the switching equipment is engaged.

*PABX.* A private automatic branch exchange, meaning an interconnected telephone system which operates in a private organization.

*Station.* A numbered telephone line associated with the PABX.

*Switch hook.* A switch which connects or disconnects a telephone instrument from the telephone line. It is activated by the telephone's carriage buttons or by the cradle holding the receiver.

The following list, in alphabetical order, is a typical sampling of features offered by the major manufacturers. The name of the feature may vary slightly in the literature of the manufacturers. Thus, I have tried to describe what the feature does for you. If you like that function, ask your vendor if he can supply the feature in your projected system. Be sure to ask what it will *cost!*

1. *Abbreviated Dialing*
   Your equipment is programmed to store frequently called numbers. You dial 2 to 4 digits, for instance, and the machine will dial the balance of the 7 to 14 digits required.

2. *Add-on Conference*
   To add a third person to a telephone conversation, the station user touches the carriage button on the telephone instrument or dials "1." The user receives a new dial tone and dials the number of the person needed for the conference. When that telephone answers, the original station user touches the carriage button of his telephone once more or dials "1." This brings his original caller back into the conversation. The three persons can now talk together. Some systems permit connecting up to

six people. No hold or line buttons are required for setting up the conference, as the mechanism required is contained in the switching apparatus.

3. *Alarm Checking and Sending*
The switching equipment sends indications of trouble to the console or to another location. An audible signal, a visual signal, or both, indicates major and minor problems.

4. *Alternate Trunk Routing*
The switching equipment can select up to five alternate routes if the route first accessed by the station user or attendant is busy.

*Attendant Busy Override* (see *Busy Override*)

5. *Attendant Busy Verification*
This permits the attendant to verify whether the station is busy, idle, or out of order.

6. *Attendant Conference* (see also *Conference Calling*)
An incoming trunk call can be connected to a number of different stations in your PABX by your console operator. If several of you in the office wish to talk simultaneously to a distant point, the operator can arrange such a conference grouping before she dials the distant number from the console.

7. *Attendant Hold*
Incoming central office calls are held at the switchboard until the operator can process them.

8. *Attendant Recall*
This permits the station user to depress the switch hook to recall the operator while connected to an incoming or outgoing trunk or to signal the operator for assistance by dialing "0" while connected to the trunk.

9. *Attendant Signaling*
A separate button is installed on the station user's instrument for signaling the operator. The operator will then call the station as soon as possible to secure instructions.

10. *Attendant Transfer*
When a call is to be transferred, the station user touches his telephone's carriage button once and dials "0," which recalls the operator so that transfer instructions can be given to her. Sometimes the system is wired so that the station user dials "1" to recall the operator.

*Automatic Call-back* (see *Call-back*)

11. *Automatic Recall*
This feature reminds the operator that attention must be given to a waiting call which has been camped on to a busy station or to a call which is waiting for the called telephone to answer. In either case, after a certain number of seconds have passed, the called number is displayed on a screen, which sometimes is an integral part of the attendant's console.

12. *Blocked Release* (see also *Sequence Call*)
A hold-and-recall feature allows the attendant to prevent automatic disconnect of an outside call if the caller makes such a request. When the call has been completed, the attendant is signaled automatically to reenter the trunk call and extend it to another telephone. This feature is helpful when the caller wishes to talk to several different departments on the same connection.

13. *Busy Lamp Field*
A panel with telephone numbers embossed on a glass screen is illuminated when a specific line is busy. This avoids the need for dialing or key-pulsing the line to test for busy.

14. *Busy Override*
If you are talking on your line, the operator can interrupt your conversation to announce that another call is waiting. A tone announces that she is entering your conversation.

15. *Call-back*
When the station user dials another station in his own telephone system and finds it busy, he dials a single digit immediately after hearing the busy signal. He replaces his handset. When the called station becomes idle, the calling station will ring. If the caller is present and picks up his telephone, the mechanism rings the telephone which was formerly busy, and the two persons are connected. In some systems, the call-back request is canceled automatically after 15 minutes if both stations have not become simultaneously idle within that time. However, the system permits only a limited number of call-backs to be stored at one time.

16. *Call Blocking*
The attendant may prevent any station or group of stations from receiving incoming calls. When activated by the attendant, the feature diverts calls to the attendant or to a recorded announcement or tone.

17. *Call Forwarding*
The station user can program the switching equipment so that it will forward an incoming call to a selected extension within the PABX system, or the call may be forwarded by an outgoing trunk to a distant location. Usually the programming is done by dialing a code.

18. *Call Pickup*
The station user sees or hears an indication that a telephone is ringing at some location within his particular group of telephones. By dialing a special code, the user can answer the ringing telephone. This feature eliminates the need for multi-button instruments in a particular department.

19. *Call Splitting*
The attendant can converse privately with an outside caller or can consult privately with the requested station without losing the incoming call.

20. *Call-waiting Display*
This is a visual indication of trunk calls waiting to be answered. One lamp may indicate that any number of calls are waiting, or there may be an individual lamp for each incoming trunk.

21. *Camp-on Busy*
An incoming call to a busy station is stored until the line is free. It is then connected automatically without further handling by the attendant.

*Centralized Dictation Service* (see *Dictation Access)*

22. *Class of Service* (see also *Line Class Marking*)
The switching equipment can be programmed to permit the station user to dial calls throughout the world, or the user may be subject to various kinds of restrictions on dialing. In some systems, class of service can be assigned on an individual basis. In other systems, users of a defined class of service are classified into a group of 100 numbers capacity.

23. *Code Call*
The attendant and station users can dial an access code and a called-party code to activate a visual or an audible signaling device. When the called person recognizes his code, he can answer by dialing his answer code from any station.

24. *Conference Calling* (see also *Attendant Conference*)
Station users and one or more outside callers can talk together on the conference circuit. Station users participate by dialing a designated number.

25. *Consultation-Hold*
To consult privately with another desired station while holding the original call, the station user touches the carriage button on the telephone instrument or dials "1." The user receives a new dial tone and dials the number of the wanted station. When the consultation is finished, the station user touches the carriage button again or dials "1" again. This brings the original caller back to the conversation. This feature is incorporated in the switching mechanism, so the instrument itself does not require the use of a hold button or line buttons for consultation-hold calls.

26. *Daytime Transfer*
If the attendant fails to answer a trunk call within a prearranged time, the call will also automatically signal one or more secondary preassigned answering stations.

*Dial Inquiry* (see *Consultation-Hold*)

*Dial Transfer* (see *Add-on Conference*)

*Dial-up Conference* (see *Conference Calling*)

27. *Dictation Access*
Special lines are provided for terminating calls in dictation recording equipment, which is then activated and controlled by single- or multiple-digit dialing.

28. *Direct Inward Dialing (DID)*
PABX stations are assigned public directory numbers so that they can be dialed directly from the public telephone network without operator assistance.

29. *Direct Outward Dialing (DOD)*
A station is given access to outside central office trunks through dialing a single or multiple digit.

30. *Direct Station Selection (DSS)*
The attendant can complete incoming calls by pushing a button designated for a particular internal station line. The button is located in matrix format on a panel associated with the console.

31. *Distinctive Ringing*
Two types of station ringing are provided to indicate whether the incoming call originates inside or outside the PABX system.

32. *Do Not Disturb*
An individual switch for each station can be activated to divert that station's calls to the console operator. Or the operator can manipulate a key and then dial the station number. This action

diverts incoming calls for that station to the console, and a lamp may also glow at the telephone so treated. The station can still make outgoing calls during the diverted period.

33. *Elapsed Time Message Metering*
A meter counts the number of calls dialed on an individual station line. The meter may be programmed to count the call after only 30 to 40 seconds have elapsed to help ensure that "busy" or "no answer" calls are not recorded.

34. *Emergency Reporting System*
An access code permits any station user to reach a group of emergency numbers which ring continuously until answered. Provision can be made to override busy lines.

35. *Executive Right of Way*
If a person wishes to override an ongoing conversation, an additional digit is dialed after the busy signal is heard. A warning tone, which is heard by all participants, indicates the third party's entrance into the conversation.

36. *Extendable Information Trunk*
The station user can place an inside or outside call by dialing the attendant, remaining on the line while the attendant completes the call or extends the dial tone to the caller. Thus no call-back is necessary when the attendant is ready with the call.

*Extension-to-Extension Dialing* (see *Station-to-Station Dialing*)

37. *Fault Indication*
Visual indications on the console show problems in the system.

38. *Flexible Number Service*
This feature permits the station number to correspond to a room number of an office, a hospital, or a hotel.

39. *Fully Restricted Service*
The station user is given a telephone line which has its own individual number. The line may be called by any other telephone in the PABX system, and it may call any other line. However, it may not receive calls from the telco public switched network, nor may it make such calls. The line does not have access to the operator.

40. *Group Hunting*
When a called station is busy, the equipment selects a free line in a predetermined sequence, and that line then accepts the incoming call.

41. *Howler Alarm*
If a station user leaves his instrument off the hook, a strong signal is heard to alert him to take action.

42. *Incoming Call Distribution*
Incoming calls are distributed equally to idle attendant consoles.

43. *Instant Ringing*
The called telephone rings instantly after the last digit is dialed.

44. *Intercept*
A call to a vacant number is routed to the attendant, to a recorded announcement, or to a busy signal.

*Key Call Operation* (see *Tone Dialing*)

45. *Line Class Marking* (see also *Class of Service*)
Individual extensions may be restricted from access to central office trunks, WATS, FX, or tie-line trunks.

*Line Ended Paging* (see *Voice Paging*)

46. *Line Load Control*
When the telephone system becomes exceptionally busy, this feature will block selected stations from making outgoing calls, thus helping ensure that incoming calls can be answered. The system automatically returns to normal operation when the peak load has decreased.

47. *Line Lockout*
This feature releases the common control equipment when a station delays in dialing after seizing a circuit or when a station develops line trouble. Thus, an off-hook station does not occupy equipment needed by others.

48. *Links*
These circuits, sometimes called junctors, provide paths for internal calls between stations. In some systems, an "optional auxiliary link circuit" is required for station access to outside trunks. The term "switch grading" refers to the number of links provided.

49. *Local-Call Registers*
The register indicates the number of calls made by a station which can directly dial local calls without attendant handling. This feature is used frequently in hotel systems.

50. *Lockout*
The attendant cannot reenter a connected station call unless

recalled by the station user. The feature provides additional assurance of privacy.

51. *Malicious Call Hold*
The called user has control of the connection so that the calling person cannot disconnect the circuit by hanging up. Some systems provide a teletypewriter display of the number of the calling telephone, or there may be trunk-route information displayed concerning the incoming call.

52. *Meet-Me Conference*
The station user or attendant sets up a conference connection by calling all involved stations, asking that they dial an assigned conference code at a predetermined time. Arrangements are also provided for connecting an outside trunk call to the conference.

*Message Metering* (see *Local-Call Registers*)

53. *Message-waiting Lamp*
This lamp is installed on the telephone instrument or in an adjacent location. A lighted lamp indicates that a message is waiting at a central location.

*Multiline Reporting System* (see *Emergency Reporting System*)

*Multi-PBX Service* (see *Tenant Service*)

54. *Music or Message on Hold*
Music or a recorded message can be provided for the waiting caller.

55. *Night Answer—Predetermined*
The attendant turns a switch or key when closing the switchboard. Then preassigned trunks are routed to preselected stations so that persons calling those trunks will be routed to a specific station after hours.

56. *Night Answer—Universal*
An incoming central office trunk call activates visual or audible signaling devices located in strategic places. Any station user with an unrestricted telephone can answer the incoming trunk call by dialing a single-digit code.

57. *Nonconsecutive Number Hunting*
Hunting is provided in a predetermined pattern rather than a sequential pattern.

58. *Number Display*
A panel on the attendant's console visually identifies the number of the station which is calling the operator. The final numeral appearing on the console identifies the class of service

to which the station user is entitled, so the operator will know if the request for service can be honored.

*Number Group* (see *Random Number Assignment*)

*Number Translation* (see *Flexible Number Service*)

*Power Failure Transfer* (see *Reserve Power*)

59. *Random Number Assignment*
Station numbers need not correspond directly with the line terminal numbers. A person moving to another location within the same dial system may take his telephone number with him without the usual rewiring being required.

60. *Reserve Power*
An alternate, independent source of power is furnished to operate the system when commercial power fails.
*Reset Call* (see *Step Calling*)

61. *Sequence Call* (see also *Blocked Release*)
If an incoming caller wishes to talk to a number of persons in sequence, the operator can activate this feature so that the call is automatically returned to the switchboard as each conversation terminates.

62. *Service Call Dialing*
Single-digit codes are assigned to frequently called service functions to save dialing time.

63. *Station-to-Station Dialing*
Stations within the system can dial other stations without the assistance of the attendant.

64. *Station Transfer*
To transfer an incoming or outgoing trunk call without operator assistance, the station user touches the carriage button of his telephone or dials "1." The user receives a new dial tone and dials the number of the station to which the call is to be transferred. If the called station answers, and if the person answering can act on the customer's request, the answering person stays on the line. The original station user touches his carriage button or dials "1." This brings his original caller back into the conversation. The original called person explains that the call is being transferred to a specific person. The original called person hangs up, and the calling person is now talking to the second person dialed.

65. *Step Calling*
If the called station is busy, another station in the same group

of stations may be called by dialing the last digit of that station number.

66. *Tandem Trunking* (see also *Tie-Line to Tie-Line Traffic*)
The attendant may connect an incoming trunk to an outgoing trunk so that the outside caller may be given access to a needed circuit, such as a WATS line or tie line.

67. *Tenant Service*
Several organizations may share a PABX system, each having its own attendant console, lines, and trunks.

*Termination of All Calls* (see *Daytime Transfer*)

*Tie-Line Service* (see *WATS Service*)

68. *Tie-Line to Tie-Line Traffic* (see also *Tandem Trunking*)
A distant station user whose call is received on a tie line can be connected to another tie line through the PABX console.

69. *Tie Trunks*
A station user reaches another PABX telephone system either through dialing one or more digits to gain access to the tie trunk or through dialing the console operator and requesting the tie trunk.

70. *Toll Diversion*
The equipment is programmed to analyze the numbers dialed by the station user. If the user dials "9" and then a toll access code such as "1," the call is diverted to the PABX operator or to a busy tone. If the station user dials "9" and then dials "0" to reach the utility's operator, the call is also diverted. In areas where no access code is needed for DDD, the equipment will analyze the first three digits dialed and divert the call if those digits do not identify a local exchange.

*Toll Restriction* (see *Line Class Marking*)

71. *Toll Ticketing*
A printout furnishes details on calls made by station users. The equipment can be designed to list toll calls, calls over dedicated facilities such as WATS and tie lines, and calls over local trunks subject to message unit charges. The station number, the area code and number of the called telephone, the time of connect and disconnect, and the date usually appear on the printout. The outgoing trunk may be identified. Cost must be allocated by applying a software procedure to the printout data.

72. *Tone Dialing*
Push-button dialing may be done on selected telephones or on all telephones within the system.

*Touch Calling* (see *Tone Dialing*)

73. *Traffic Metering*
All links and trunks are arranged to provide a continuous indication of occupancy so that external portable recording devices can function when connected to the links and trunks.

*Trunk Restriction* (see *Line Class Marking*)

74. *Trunk-to-Trunk Consultation*
If the station user is connected to any kind of outgoing trunk, including WATS, tie lines, and FX, the user can leave that trunk to gain access to a second outside trunk for consultation, but not for conferencing.

75. *Two-Party Service*
Two extension numbers can be assigned to a single station line. Each number can be rung separately.

76. *Voice Paging*
The attendant and/or station users have dial access to a paging system.

77. *Wake-up Service*
Wake-up requests are arranged in sequence, so when the station is answered a recorded message can be given automatically, or the answered stations are connected simultaneously to the attendant.

78. *WATS Service*
Dial access codes are provided to permit station users to secure WATS lines without operator assistance.

## INTERCONNECTION OF SUPPLEMENTAL TELECOMMUNICATIONS EQUIPMENT

Now that the FCC has authorized the interconnection of equipment, manufacturers of telecommunications equipment have mounted extensive advertising programs to stimulate the lease or sale of their particular merchandise. The general public is being urged to acquire automatic telephone answering equipment and automatic dialers. The small business manager is learning about inexpensive equipment which will hold a call on one line while a secretary dials the manager at his residence. When buttons are depressed on an instrument which is connected to the six-button key telephone, the incoming call may be bridged to the manager's residence without a second call being made. More expensive equipment will automatically accept an incoming call, hold it, and dial any number which has been preset by the businessman, forwarding the call automatically to that number (see Chapter 11).

In Chapter 3 on WATS, you learned about interconnected devices which provide automatic toll-ticketing information so that toll charges can be allocated to the originating station or even to the person who dialed the call at the originating station. Many residential telephone instruments are being sold to replace extension telephones previously rented from telco. In the field of data communication and administrative written message service, it is quite common for the customer to provide his own teleprinter terminal, which has been leased or purchased from a private vendor.

If the interconnected supplemental equipment is wired into the utility's system, an interface device is required under tariffs in effect during 1973. As was mentioned earlier in this chapter, those tariffs may be supplanted by a program which certifies such equipment as being harmless to the utility's network. If you are considering interconnecting supplemental equipment, be sure to determine the monthly rental charge which telco will make for the required interface. In addition, telco may make a charge for installing different parts in the interconnected equipment to minimize or eliminate possible problems which might arise from its operation.

On the other hand, if an acoustic coupler is used by the interconnected device, no interface is required. Installation is simplified, and the user may find that acoustically coupled devices indeed provide useful services. Telephone answering equipment has been discussed in Chapter 11. Another type of acoustically coupled equipment which is being used widely is the facsimile transmitter. Because the use of facsimile has grown very rapidly, it seems advisable to give you some basic information on facsimile transmission.

## FACSIMILE TRANSMISSION

A facsimile transmitter permits the user to transmit a copy of a document from one point to another point which is equipped with a compatible facsimile machine. The use of facsimile is justified when the speed requirement will not permit the use of mail, or when the document cannot be copied and transmitted by Telex, TWX, or

Mailgram, or when the information cannot be read to the recipient over toll, WATS, FX, or tie-line circuits.

Facsimile transmission is used to transmit engineering drawings, graphic material, complex statistical tabulations, legal documents, and similar material, some of which would require many hours of labor to create tapes for transmission on Western Union domestic service or on private line networks. In general, facsimile equipment can satisfactorily reproduce material printed in type as small as 4-point.

## The Equipment

Facsimile equipment consists of a scanning device and a transmitter. The scanner looks at the document with a bright light. A photoelectric cell interprets the amount of reflected light. The image is converted into a digital electric signal. The signal is picked up by an acoustically coupled interface device, which may be an integral part of the machine, and is transmitted over any telephone line of voice or data grade. The sender dials the number of the distant facsimile machine, which is a duplicate of the sending machine, or another manufacturer's machine which is compatible with the sending machine. The document is re-created at the receiving point. The two most common methods of re-creating the image are electrically burning an impression on the paper or pressure tracing of the image through carbon paper.

## Speed and Resolution

The quality of the document received at the distant end depends on the scanning speed of the originating machine, expressed in lines per inch. The pinpoint of light from the photoelectric cell travels along the copy to be transmitted. The closer and the more numerous the scanning lines, the better will be the resolution or clarity of the reproduced image. For instance, one manufacturer provides a facsimile transmitter which can transmit an 8½ by 11-inch document in six minutes. That machine has a scanning speed of 96 lines per inch. To speed service and to cut the cost of long distance toll charges or WATS usage charges, the facsimile user is always looking for a faster machine. The same manufacturer offers a machine which transmits the same 8½ by 11-inch document in four minutes, but the scanning speed drops to 64 lines per inch, so the reproduced document is less sharp.

A competing manufacturer advertises a six-minute machine at somewhat less scanning speed, 88 lines per inch, and he also offers a three-minute machine at that same scanning speed. One manufacturer announced that a machine to be available in 1974 would cut the time to one minute.

Manufacturers in this field are highly competitive. The potential user of facsimile must compare the rental or purchase cost of the unit with the speed and clarity of transmission. It is quite possible that the higher rental or purchase cost of a speedier machine may be amortized quickly by savings in toll or WATS usage charges. The potential user should talk to all the major firms offering equipment, test the equipment in actual use, and then make an informed decision.

In 1973 manufacturers were offering dependable slow-speed equipment with high resolution. This type of equipment is used for transmitting weather maps and charts, for criminal identification, and for military, news, and Wirephoto applications. Such slow-speed equipment may be quite useful in intracity or intrabuilding applications, such as forwarding copies of international Telex and cable messages from the international department to the dispatching message center. Most of the facsimile market utilizes higher-speed equipment with less sharp resolution. Much of the material transmitted is typed or drawn copy, and the main requirement for transmission is acceptable legibility at the distant point.

Recent studies indicate that facsimile usage is growing rapidly. One study indicated that a half-million machines would be operating by 1975.

## Cost

The least expensive six-minute machine was renting at $50 per terminal in 1973, and it offered 96 lines per inch resolution. A three-minute machine offering 88 lines per inch resolution was renting at $88 monthly.

## Additional Features of Facsimile Equipment

**Compatibility.** In the early development of the industry, manufacturers tried to make their transmitters noncompatible with competing equipment so that the user would standardize on the equipment of one manufacturer. More recently, all major manufacturers make at least one transmitter which is fully compatible with competitive equipment.

**Duplex Operation.** Most equipment operates on a full-duplex basis, which means that the equipment can send and receive simultaneously, thus cutting down transmission charges.

**Unattended Operation.** All manufacturers offer this feature. In one machine, up to 75 documents can be loaded at one time, and the machine will feed them to the scanner automatically. Such machines are equipped with roll paper feed, eliminating the need for constant attention by the operator. The receiving machine can shut itself off automatically, also.

**Simultaneous Broadcasts.** In 1973 one manufacturer advertised that a single document could be broadcast simultaneously to as many as 100 receiving facsimile machines.

**Directory Service.** One manufacturer offered its customers a free directory service through calling an inward WATS "800" number.

**Access to Telex-TWX.** An organization announced a new service in 1973 named "T-FAX," which permits any user of facsimile equipment to have a facsimile transmission routed to the organization's message center via an inward WATS "800" number. At the center the typed copy is converted to tape and dispatched to the requested Telex or TWX number. For details, see Chapter 14.

## THE FUTURE OF INTERCONNECTION

If you are considering the acquisition of an interconnected telephone system, or if you wish to obtain some apparatus to be interconnected to the utility's network, you must be satisfied that your investment will be justified by the quality of the equipment you acquire and, equally important, that you will be allowed to make the interconnection. You must be scrupulously careful to pick a manufacturer which is strong financially and which has a proven record of quality production. The weaker, newer organizations may fail, or they may be absorbed by the stronger companies. Be sure that the equipment you wish to get will have maintenance and repair parts support from an organization of unquestioned strength and integrity.

As you have learned in this chapter, it is indeed possible to acquire interconnected equipment of excellent quality and proven performance. Further, you have learned that the principle of interconnection is not new, but has existed for many years. In my judgment, the courts will sustain the decision of the FCC to permit interconnection on a broad scale. I also believe that the state regulatory commissions will permit interconnection within their specific jurisdictions.

As time goes on, interconnected equipment will make greater use of electronic switching. Stored programs will be available within a computer which will be an integral part of the electronic switch. For instance, these programs will permit telephone numbers to be changed by a teleprinter. Such systems will likewise include video capabilities so that executives will be able to see each other while talking.

Despite the many cautions I have emphasized in previous pages, I believe that any user of a large telephone system should investigate the service improvements and improved return on investment which may come about from an interconnected telephone system acquisition. The intensive planning and investigation required may be well worth its cost. Further, the cost of the interconnected system may become more attractive as utility rates continue to increase. It seems quite possible that as interconnect vendors increase their share of the market, the utilities will concentrate their efforts on furnishing the facilities to which the interconnected equipment must be attached, improving the quality of those facilities, and adjusting rental rates to reflect those improvements.

# CHAPTER THIRTEEN

# USING THE POSTAL SERVICE TO SUPPLEMENT TELECOMMUNICATIONS

What is the United States Postal Service doing for the administrator of telecommunications services? Is the mail service really improving? Is it dependable? Is it fast enough to justify diverting expensive written wire communications to the mails?

Uncle Sam's Postal Service indeed offers new, helpful services to both the private citizen and the business user of the mails. Some of the newer services have not been widely publicized. If you will study them, you may find that you can use mail services for written material which is now moving over expensive domestic or international telecommunications channels.

## AIRMAIL

In 1972, the Postal Service published a bulletin entitled "Airmail—The Preferred Service." The introduction states:

> When you consider that the world's biggest communications system—740,000 employees, 44,000 postal locations and a multibillion-dollar plant—gives first preference every day in every way to moving airmail, the airmail letter is one of the world's largest bargains.

The Postal Service has established a one-day delivery standard for airmail addressed to principal cities within 600 airline miles from the originating principal city. This service also has been extended to designated cities up to 1,200 airline miles from the originating city. Your postmaster can identify those cities. A two-day airmail delivery standard applies to all other cities within the continental United States.

To be eligible for delivery on the following business day, airmail must be zip-coded and be deposited before 4 P.M. in a white-topped mailbox or in the designated box or slot in a post office. Airmail from the special boxes or slots is placed in a distinctive orange sack. The collecting carrier may place those sacks on a vehicle that travels directly to an airport branch post office, where the mail receives an "AMF" (Airmail Field) postmark. Generally, airmail is given preference in all mail handling.

At destination, airmail letters take precedence over first-class letters in being sorted and distributed to the mail carriers. The Postal Service says: "The carrier leaving on his delivery route in early morning may very well have the airmail letter in his satchel for delivery while the first-class letter mailed at the same time is still en route."

In late 1973, the Postal Service began to move first-class mail by air on a "space committed" basis rather than on the former "space available" basis. This action was taken to meet the new delivery standards applicable to first-class mail. Thus, first-class mail may sometimes arrive as fast as airmail, but the two-day maximum airmail delivery standard may still justify your payment of the premium rate.

The Postal Service has worked out cooperative programs with foreign governments for expediting international airmail. Two-day airmail service is now possible between the United States and many points in Great Britain and other European countries. Airmail may be coupled with "Expres," the international term for special delivery, for even faster handling. On request, the Postal Service furnishes special international "Expres" stickers for foreign mail. In addition, the Postal Service sells aerograms (air-letter sheets) for international service. In 1973, the aerogram sold for 15 cents and provided airmail service to any country. No enclosure is permitted, nor can it be sealed with stickers or tape. Because the aerogram may reach its destination within a day after an overnight cable would have been delivered, many multinational organizations are making extensive use of this low-cost medium of international communications, particularly on weekends.

A little-known provision of the postal regulations permits the private manufacture of air-letter sheets. This regulation is found in the International Mail manual, section 415.2. A company can follow the specifications given in the instructions, apply for a permit, and submit samples for approval to:

Mail Classification Division
Finance Department, U.S. Postal Service
Washington, D.C. 20260

If you wish, before you submit copy to your printer, the Mail Classification Division will send you a sample format.

This regulation permits you to print aerograms in quantity, with your letterhead appearing on one side. Postage can be applied by meter or adhesive stamps at the time of dispatch. The Postal Service will accept the private aerogram at the regular airmail rate for Canada and Mexico and at the 15-cent rate for other countries.

As my own corporate office is in Cleveland, and as my office dispatches airmail to most of our country's principal cities, we have been experimenting with the new airmail program. I am glad to say that the delivery standards are being met between principal cities with only a few exceptions.

If this program interests you, check with your mail-room supervisor and determine whether airmail is being segregated and pouched or packaged separately from first-class mail. If you have a considerable volume of airmail, your post office will furnish you with orange airmail sacks which will receive preferred handling. Then check the time at which your airmail is dispatched from the mail room. You may wish to make a special trip to your nearest post office or white-topped airmail box to ensure that you observe the 4 P.M. cutoff time for next-day delivery.

Also, check with the manager of your international department to determine if your company's foreign offices are open on Saturday. If not, you may wish to question whether you should dispatch international overnight cables on Friday. The information could be sent in special airmail envelopes or aerograms, and delivery to many cities would be made on the following Monday. The cost reduction in changing the cable to airmail dispatch will be very substantial. You may also wish to divert Friday afternoon dispatches of international Telex messages to airmail.

If your company permits departmental secretaries to enclose letters in special airmail envelopes, you may occasionally wish to check the destination of those letters. Your postmaster can tell you the cities which are reached overnight by surface transportation. You can then instruct the departments which originate this traffic not to use airmail to those nearby points.

## SPECIAL DELIVERY

This service has become less useful because delivery frequency has been reduced. Also, the cost has increased substantially; a special-delivery stamp sold at 60 cents in 1973. Many organizations report that regular airmail is delivered earlier than mail dispatched as airmail special delivery or as special-delivery first-class mail. Special delivery is sometimes used for prestige reasons. The recipient knows that the sender wants him to receive the mail at the earliest possible time. Also, special delivery may ensure delivery on Saturdays or Sundays, when regular carrier service is reduced or eliminated. Special-delivery mail received at a post office before 8 P.M. should be delivered that day.

## THE WHITE STAR SYSTEM

The Postal Service has been curtailing the frequency of collections from street mailboxes. Mail deposited in boxes located in residential areas may be picked up by the carrier as he delivers mail. In most residential areas, the final collection is made in early afternoon.

To help the public identify boxes where later collections are made, the Postal Service decided to place one or two white stars on certain mailboxes. A single star indicates that mail in the box will be collected around 5 P.M. Two stars indicate that a later collection will be made between 6:30 and 8 P.M.

In general, one-star boxes are located in business or industrial areas and on main traffic arteries in residential and shopping center districts. Two-star boxes normally are located in downtown areas and outside of branch post offices. Be sure that your mail is being deposited in boxes which will be collected later that day. Secretaries who personally mail last-minute letters may be grateful if you acquaint them with the white star system.

## NEXT-DAY DELIVERY

The Postal Service is gradually establishing a geographic area surrounding each post office where first-class mail will be delivered on the following day. New stickers are being affixed to mailboxes showing zip-code areas where next-day delivery can be made. Your postmaster can give you a list of cities surrounding your city where such expedited service can occur if you deposit your mail in time for collection by 5 P.M. In general, the Postal Service is working toward a standard which will provide next-day delivery within your *own* city if mail is deposited in a box which is collected by 5 P.M. or earlier.

## SATURDAY COLLECTIONS

If members of your organization work on Saturdays, be sure that you are familiar with the collection schedule for the mailbox or mail chute which you use on Saturdays. In some office buildings collections on Saturday are made only in the morning, and Saturday afternoon collection of mail has been reduced. White star boxes should be collected around 5 P.M., but often collections are scheduled around 2 P.M. Your mailbox or mail-chute collection schedule should be examined so that you can give your Saturday workers proper instructions. If people make a special effort to complete a project on a Saturday afternoon so that a presentation can be mailed for Monday delivery to a customer, you'll want to make sure that their mail is indeed dispatched on Saturday.

## SPECIAL POSTAL SERVICES FOR THE BUSINESSMAN

In 1972, the Postal Service announced a group of specialized postal services for business use. A high-speed delivery network was created, linking more than 50 major cities in the United States and England. Western Union and the Postal Service collaborated on the development of the Mailgram, a letter-telegram which is originated by telephone, Telex, TWX, or Info-Com and is delivered the next day by the postal carrier. The Mailgram is described fully in Chapter 14. Here are some facts on the other services.

### Express Mail Service

This service links most major United States cities. If desired, you can schedule your specific delivery requirements with your post office on a contractual basis. A noncontractual "downtown-to-downtown" service is available. Here are some variations of express mail service. You can secure rates and further facts from your local postmaster.

**Door-to-Door.** On a regularly scheduled basis, a postal driver arrives at your place of business at a designated time to pick up your urgent dispatch. It is taken directly to the airport, where a special agent transfers your dispatch to the next aircraft departing for your destination city. Upon arrival at destination, your mail is handed by a special agent to a second driver, who makes delivery to your addressee at a designated time.

**Your Door-to-Destination Airport.** This procedure is the same as door-to-door, except that your recipient must pick up the mail at the destination city's airport postal station.

**Originating Airport to Addressee.** You furnish your own driver or messenger, who takes your mail to your airport postal station. A special agent places your dispatch on the next available aircraft departing to your destination city. When the dispatch arrives there, an agent transfers it to a postal driver for delivery to the addressee within a designated time.

**Airport-to-Airport.** You arrange to dispatch your mail to the airport designated by the post office. An agent accepts your mail and places it on the next aircraft departing to the destination city. Your recipient picks up the mail at the destination city airport postal station.

**Downtown-to-Downtown.** Certain post offices in your city are designated as pickup and delivery points. You take your letter or parcel mail to one of those points prior to 5 P.M. The dispatch is placed on the fastest mode of transportation to the destination city. Your addressee picks up his express mail at the downtown post office after 10 A.M. the next morning. The Postal Service guarantees that the mail will be delivered within 24 hours. If not, your postage payment is refunded. The recipient signs a receipt to provide proof of delivery. Saturday service is provided in certain cities. In 1973, rates for delivery at the destination post office ranged from $1.50 (½ pound) to $22 (40 pounds), depending on weight and distance. Optional delivery to the addressee's office or residence costs an additional $4 to $5. Express mail can contain merchandise as well as letters, computer runs, and other paperwork.

**United States to England.** The Postal Service has linked its express mail service with the Datapost service of the British post office. On a contractual basis, mail can be expedited to all cities in the United Kingdom. The service is offered on a door-to-door or airport-to-addressee basis.

The Postal Service publishes a group of special bulletins which describe the special services I have mentioned in this chapter. Your local postmaster can furnish copies.

## Other Postal Services

Since the Postal Service became a corporation under private management, there has been extensive modernization of facilities and services. Every user of telecommunications services should be alert constantly to the possibility of reducing the cost of domestic and international written wire communication by utilizing the modern services which the Postal Service has introduced.

The Service may be able to tailor-make a specific service to your needs. For instance, one organization has extensive South American operations. Airmail service was found to be very slow. It was difficult to secure lines from the government telecommunications administrations to utilize international Telex. Cable deliveries were delayed. Working jointly with the Postal Service, this organization reserved cargo space on an aircraft which departed daily for the principal South American city in which business was conducted. A postal worker picked up the mail dispatch in the originating city and dispatched it via express mail service to New York's Kennedy International Airport. Another postal worker in New York placed the dispatch on the aircraft. At destination, a third postal worker picked up the dispatch and delivered it to the corporation's principal South American office. You may find that your local postmaster, or his superiors in Washington, can design similar helpful services for you.

# CHAPTER FOURTEEN

# WESTERN UNION DOMESTIC SERVICES

In many business situations, there is a less expensive alternative to traditional voice communications. Written messages, dispatched over Western Union's inexpensive telegraph-type circuits, can cut the cost of many telephone calls by as much as 75 percent.

This chapter describes the *written* message services provided to the public by Western Union, a major common carrier. Telex and TWX are two Western Union services which are widely used by organizations throughout the United States. A newer service, Mailgram, is a joint undertaking of the United States Postal Service and Western Union. It has been very successful. Mailgrams are rapidly supplementing venerable full-rate and overnight telegrams.

Before discussing Western Union's services in detail, let's talk about some ideas which will help you convert expensive voice communications to a written message which can be dispatched over an inexpensive telegraph-type circuit.

## TYPES OF TELEPHONE CONVERSATIONS

If you stop to consider the subject for a moment, you'll realize that all voice conversations can be classified broadly into three types.

### The Discussion Conversation

The greatest majority of voice conversations are truly required because the persons involved must engage in a discussion. Sometimes there is a prepared agenda, as suggested in Chapter 1. Otherwise, each person has a mental outline of the subject to be discussed. The telephone call is a quick way of answering questions which have arisen. Then prompt action can be taken. Particularly when the matter is urgent, such a telephone call is necessary and desirable.

### The Request for Information

There may be times when we do not need to discuss a subject with the distant person—we only need to secure information from him. For instance, we need a dimension of a machine tool part which we have been asked to manufacture. Often, the person we telephone must put our call on "hold" while he goes to his file to secure the facts we need. Or he may ask us to call back later when he has accumulated the information.

### The Conveying of One-Way Information

Your customer is very anxious to know whether or not you have shipped his order for a carload of lumber. He has been calling you daily to ask about his order. Your shipping department foreman announces that the car has been loaded and will be moving out that afternoon. You are delighted. You reach for the telephone to call your customer and relieve his anxiety. That call may be worth its cost because of the public relations value of your cordial voice conveying the good news. On the other hand, your customer is also a friendly person, and after you've given him the car number and the carrier's name, he'll be reluctant to be abrupt when you've been so thoughtful, so your call will run overtime until the social pleasantries are finished. Ask yourself: "Could I have written a cordial message of 15 to 30 words which would have conveyed the essence of both the facts and my enthusiasm?" From New York to San Francisco, a three-minute DDD call will cost $1.45 plus federal tax. A 30-word Telex message between the two cities would be billed to you at 30 cents. That's a cost reduction of 79 percent. Your time would have been saved, too.

## USING WRITTEN MESSAGES AS SUBSTITUTES FOR VOICE CONVERSATIONS

As you can see, the person who is trying to reduce cost and save executive time has a real opportunity to educate his telecommunications users about the economy of the written wire message services. If you were to record a typical telephone conversation, you'd find that 25 to 50 percent of the words spoken are not essential to the delivering of the actual message. Any financial officer concerned with the rising cost of telephone service will welcome any suggestion you can make on methods of converting such conversation to a less expensive channel. Your job is to explain the three basic types of calls to your users. Give them cost comparisons of Telex, TWX, and Mailgram rates with DDD, WATS, FX, or tie-line rates to the various points you call regularly. Be sure to stress the *percentage* cost reductions for messages of typical length. In short, suggest that the user *THINK* before he picks up the phone.

Changing a person's habit pattern is difficult. But in this case, very substantial cost reductions will come about from the conversion of telephone calls to written wire messages. The intangible saving in executive time by eliminating the inevitable social pleasantries should also be stressed as a "plus" selling factor when you talk with your people.

Some of your users will remind you that it costs a great deal of money to write a wire message. They assume that it is a lot less expensive to pick up the telephone. But it's going to be hard for them to argue with your statistics if you put an accurate value on the time consumed in the lengthy conversation and then price out the cost of the telephone call. It doesn't really take very long to write or type a 15- to 30-word message. A secretary can do it quickly if the executive is unwilling to write it himself. The message is placed on an appropriate form and is telephoned or delivered to the message center. There a skilled supervisor scans the message for clarity, edits it if necessary, and sees that it is dispatched by the least expensive way consistent with the delivery time and date indicated on the form by the originator.

Now let me discuss Telex and TWX services which are available from Western Union. I will give the important specifications first and then discuss the services.

## WESTERN UNION TELEX

The word "Telex" is an acronym for "teleprinter exchange." Telex is used throughout the world as a fast, inexpensive method of written communication.

*Speed:* 66.67 words per minute.

*Number of subscribers:* United States, 49,000; Canada, 21,000; Mexico, 6,000; Alaska, 200; worldwide, 260,000.

**Figure 1.** Model 32 Telex automatic-send-receive teleprinter. *(Western Union)*

*Equipment used:* Teleprinter machine with three-row keyboard.

*Basic types of equipment available:* Automatic-Send-Receive (ASR) for tape operation (Figure 1); Key-Send-Receive (KSR) for manual operation; Receive only (RO). Units operate unattended.

*Type of tape used:* Five-level Baudot code.

*Cost of transmission:* At 1973 rates, from 17½ to 60 cents per minute, depending on distance. There is no minimum time charge.

*Availability of facilities:* There are Telex exchanges in all principal cities. If a particular locality does not have a Telex exchange, a mileage charge applies from your office to a point 10 miles distant from the city limits of the closest Telex exchange city.

*Geographic coverage:* Direct transmission to Telex machines in Canada, Alaska, and Mexico. Connection via international carriers with international Telex machines abroad.

## WESTERN UNION TWX

The initials "TWX" stand for the words "teletypewriter exchange." TWX was formerly offered as a service of AT&T. Western Union bought the service in 1971 and now operates and administers it.

*Speed:* 100 words per minute.

*Number of subscribers:* United States, 48,000; Canada, 7,000.

*Equipment used:* Teleprinter machine with four-row keyboard.

*Basic types of equipment available:* ASR for tape operation (Figure 2); KSR for manual operation; RO. Units operate unattended.

*Type of tape used:* Eight-level ASCII code.

*Cost of transmission:* At 1973 rates, from 20 to 70 cents per minute, depending on distance. Charges are calculated in full minutes; a fraction of a minute is charged as a full minute.

*Availability of facilities:* TWX is available without mileage charges in almost every location.

*Geographic coverage:* Direct transmission to TWX machines in the United States and Canada. Connection via international carriers to international Telex machines abroad.

## PERIPHERAL SERVICES AVAILABLE TO TELEX AND TWX USERS

Both services have access to the Western Union computers, which offer Teleprinter Computer Service (TCS). The computer will accept the following requests for service:

**Figure 2.** Model 33 TWX automatic-send-receive teleprinter. *(Western Union)*

1. *Telex to Telex:* If a large volume of messages is to be transmitted to other Telex users, the computer will accept the messages and do the dialing, thus relieving the originating organization of the dialing responsibility. Also, this service is very helpful when your operator encounters a busy machine on a directly dialed Telex connection. The computer will try the busy number 10 times before returning the message to the originating point as being undeliverable.

2. *TWX to TWX:* A comparable service for TWX users.

3. *Telex to TWX:* If the recipient does not have Telex equipment but the Western Union directory indicates that TWX equipment is available at destination, the computer will accept the Telex message, convert it to TWX language and speed, and deliver it to the indicated TWX machine.

4. *TWX to Telex:* A comparable service for TWX users.

5. *Multiple address service:* By following a specific format, the dispatching operator can ask the computer to send a specific message to different addresses. The computer will accept a request for TWX, Telex, Mailgram, telegram, or night letter services. Up to 50 identical-text messages may be sent on a single computer connection.

6. *Mailgram service:* This service will be described in detail later in this chapter. Single- or multiple-address messages can be routed to the computer.

7. *Collect messages:* The computer will send a collect TWX or Telex message to points within the United States.

8. *Computer telegram service:* A full-rate telegram or night letter may be sent via the computer rather than telephoning the message to a local Western Union office. Rates for telegrams and night letters are less via computer than rates applicable to messages telephoned or placed on teleprinter tie lines to local Western Union offices. The TCS computer classifies telegrams and night letters as Public Message Service (PMS). If physical delivery is required, the words "physical delivery only" should be incorporated in the *address* of the message. The charge for physical delivery in 1973 via TCS was $3 within the destination city's limits. The same delivery charge applies to telegrams telephoned to Western Union's "800" number or to telegrams given to a Western Union local office.

9. *Itemized billing service:* Through a message numbering system which is incorporated in the special dispatching format used with the TCS computer, a user of TCS service can code messages numerically to indicate the cost center to which the message should be charged. The message copy retained after dispatch is used for accounting purposes.

10. *Cablegram service:* The TCS computer will accept cablegram messages.

11. *Full accounting service:* Billings to TCS users itemize each Telex, TWX, Mailgram, and PMS message dispatched during the billing period.

## RATE COMPARISONS BETWEEN TELEX AND TWX SERVICES

In a following section of this chapter, I will discuss the advantages and disadvantages of Telex and TWX under the various circumstances of usage. It will help you to understand that discussion if you study Table 1, which compares the Telex and TWX rates in effect in late 1973. At the time you read this chapter, you can ask your local Western Union office to send you the current rate schedule. Then you can make a similar rate comparison which will be up-to-date.

The full-rate telegram was the fastest method of public written wire communication prior to Telex. With careful attention to eliminating unnecessary words, an administrative message could frequently be composed within the 15-word allowance. To this day, Western Union charges for a full-rate telegram are based on a minimum of 15 words. Consequently, I have chosen a message with a 15-word text as my example of comparative cost in Table 1. The following conditions apply to this table:

1. *Length of message:* 15 words of text. If the message moves as a telegram, night letter, or Mailgram, assume that an additional 15 words will be required for address and signature. Time allowances include the time taken for automatic identification of called and calling Telex or TWX machines.

2. *Chargeable time of transmission for 15 words of text:*
Telex: $\frac{5}{10}$ minute
TWX: One minute
Telex to TCS computer: $\frac{5}{10}$ minute
TWX to TCS computer: One minute
Telex to TCS computer to Mailgram or PMS: $\frac{5}{10}$ minute
TWX to TCS computer to Mailgram or PMS: One minute

3. *Distance of transmission:* There are 12 rate areas or zones in Telex, but only 4 rates. By contrast, there are 11 rates applicable to TWX transmission, depending on the airline mileage between points. Table 1 is designed to show the minimum and maximum rates for Telex and TWX transmissions of a message with 15 words of text.

**TABLE 1 Comparative Cost of Telex and TWX**

| | TWX | | Telex | |
|---|---|---|---|---|
| | A | D | A | D |
| Telex to Telex | | | $0.0875 | $0.30 |
| TWX to TWX | $0.25 | $0.70 | | |
| TCS-Telex to Telex | | | 0.1875 | 0.43 |
| TCS-TWX to TWX | 0.35 | 0.85 | | |
| TCS-Telex to TWX | | | 0.55 | 1.15 |
| TCS-TWX to Telex | 0.85 | 1.75 | | |
| TCS-Telex to PMS—full-rate telegram | | | 3.20* | 3.20* |
| TCS-Telex to PMS—night letter | | | 2.60* | 2.60* |
| TCS-TWX to PMS—full-rate telegram | 3.80* | 3.80* | | |
| TCS-TWX to PMS—night letter | 2.50* | 2.50* | | |
| TCS-Telex to Mailgram | | | 0.84 | 1.05 |
| TCS-TWX to Mailgram | 0.83 | 1.02 | | |
| TCS-Telex to Canadian Telex | | | To any destination: 24 cents | |
| TCS-TWX to Canadian TWX† | To any destination: 70 cents | | | |
| TCS-Telex to Mexico† | | | To any destination: 33 cents | |
| TCS-Telex to Alaska† | | | To any destination: 52 cents | |

* Add $3 if physical delivery is desired within destination city limits.

† The rates quoted *include* the TCS service charge of 10 cents per message. Direct Telex-to-Telex and TWX-to-TWX service is also available, at rates which can be secured from Western Union.

Note on collect messages:
TWX to TWX: Add 25 cents per message
TCS-TWX to TWX: Add 5 cents per message
Telex to Telex: No automatic collect service available
TCS-Telex to Telex: Add 5 cents per message

The Telex rate in column A represents the rate applicable within the 17½ cent per minute Telex rate zone. The TWX rate in column A represents the cost of a transmission which travels no more than 110 miles. The Telex rate in column D represents the rate applicable between a point in the 17½ cent per minute zone and a point in the 60 cent per minute zone. The TWX rate in column D represents a transmission which travels more than 2,000 miles.

As an example, column A rates would apply to a transmission between New York and Philadelphia. Column D rates would apply to a transmission between New York and Pacific Coast points.

The source of the per-minute rates given in Table 1 is the *Product Line Reference Book, March 1973,* Western Union Telegraph Company.

## DISCUSSION OF FACTORS TO BE CONSIDERED IN CHOOSING BETWEEN TELEX AND TWX

### Length of Message

There is no *minimum charge* for a Telex transmission. Thus, if the message is transmitted in less than a minute, the sender is charged for the actual connection time. The minimum *rate* for one minute of Telex transmission in 1973 was 17½ cents. The minimum rate for one minute of TWX transmission was 20 cents. Because Telex transmits at 66.67 words per minute and TWX transmits at 100 words per minute, one might think that the cost per word of transmission would be less by TWX. This is not necessarily true, for Telex is billed in ⅒-minute increments and TWX in full-minute increments. Fractions of a minute are charged as a full minute in TWX transmissions. These differences will become clearer if you study Table 2.

**TABLE 2 TWX-Telex Cost Comparison for Messages of Various Length Sent Between New York and Philadelphia**

| Length of transmission in words | Chargeable TWX transmission time in full minutes | Cost | Chargeable Telex transmission time in ⅒ minutes | Cost |
|---|---|---|---|---|
| 15 | 1 | $0.25 | 3 | $0.0525 |
| 100 | 1 | 0.25 | 15 | 0.26 |
| 110 | 2 | 0.50 | 17 | 0.30 |
| 190 | 2 | 0.50 | 29 | 0.51 |
| 210 | 3 | 0.75 | 32 | 0.56 |

Thus, the *average length of your messages* will tell you a great deal about whether you should install Telex or TWX. If your messages are short, Telex may save you a considerable sum, even if it operates at a slower speed. If your messages are longer, they may still be more expensive via TWX unless you edit them to keep them just under the 100 words per minute speed.

If your messages must be dispatched quickly and they are short, Telex will provide you with fast, inexpensive service. The transmission cost is so low that messages can be taped and dispatched as soon as they reach the message center. However, if time permits messages to be accumulated and sent as a group to one destination, TWX may save you money because its 100 word per minute speed is 50 percent higher than Telex. But be sure to read the next section on the geographic location of the city you're calling.

## Rate Structure Differentials

Your choice between Telex and TWX service may be influenced by *the geographic location you plan to call.* Telex destinations are divided into 12 zones, but only 4 rates apply to those zones. For a call within the same zone, the lowest rate applies. A call to a contiguous zone takes the next highest rate. A call to a zone once removed takes a still higher rate. All other zones take the highest rate. In 1973, the four rates were:

| | |
|---|---|
| Rate A | 17½ cents per minute |
| Rate B | 30 cents per minute |
| Rate C | 45 cents per minute |
| Rate D | 60 cents per minute |

An an example, a message from a city in one state to a city in another state in the following list of states would be classified within Rate A:

| | |
|---|---|
| Connecticut | New Jersey |
| Delaware | New York |
| District of Columbia | Pennsylvania |
| Maine | Rhode Island |
| Maryland | Vermont |
| Massachusetts | Virginia |
| New Hampshire | West Virginia |

Now, let's compute the cost of a transmission from Boston to Pittsburgh by Telex and by TWX. As both cities are in Telex Rate A, the cost of a one-minute transmission will be 17½ cents.

TWX rates are based on airline mile distances. There are 11 rate zones, which range in cost per minute from 20 to 70 cents. A minute of TWX transmission from Boston to Pittsburgh, which is a distance of 482 airline miles, will cost 45 cents, a difference over Telex of 157 percent.

You can see how important it is to list the cities you'll be calling and then determine the rates applicable for both services. You may find that Telex service will be much less expensive, even considering its slower speed.

## Data Usage

The five-level tape produced by Telex may not be compatible with the requirements of your data processing department. For instance, if your Western Union TWX or Telex equipment is being used for transmitting orders, your data people may want you to produce a tape on *incoming* transmissions so that those tapes can serve as direct input to the computer. In this case, you may find that the eight-level tape produced by TWX will better serve your needs. However, it is possible to acquire a tape-to-tape converter which will create an eight-level tape from a five-level tape. If your messages are short, and if you do not group them together into batches for longer transmissions, Telex may still be the preferred service if you acquire the tape-to-tape converter.

Western Union offers an "alternate use" switch for its TWX service. A teleprinter used primarily for TWX can be switched to transmission over the DDD toll network or over any voice circuit such as WATS, FX, and tie lines. Thus, a single machine does double duty, and it can be utilized through appropriate data terminals for night transmission of low-speed data. Such service is likewise available for transmission to TWX users in Canada who are equipped to receive data transmission.

Many organizations are utilizing computers physically located at a distant geographic point. The time-sharing concept has gained much momentum in recent years. A teleprinter terminal is provided in your office. You use the terminal to access the distant computer. Such teleprinter terminals may be selected so that they are compatible with Western Union's TWX public network. A simple switch converts the operation from computer time-sharing to the TWX network. Conversely, a teleprinter furnished by Western Union for TWX network access can be used for computer time-sharing. The saving in machine rental or purchase cost may be a factor which influences you in a choice of TWX over Telex. Again, be sure to do your economic study on the average length of your typical message and its cost of transmission over each of the two services.

Some companies may use a private teleprinter network for intracompany communication. If the teleprinters used are compatible with TWX network characteristics, the Western Union tariff permits "customer-owned" equipment to be switched into the TWX network through appropriate interface devices. This arrangement may save the rental cost of a separate teleprinter used exclusively for TWX service.

## Industry Practice

Because TWX service was established prior to the wide use of Telex in the United States, certain industries standardized on TWX as the desired method of communication between organizations in those industries. For instance, the electronics industry makes great use of TWX. Municipal bond dealers employ TWX for trading between different investment banking organizations and customers such as bank trust investment departments. Newspapers use TWX widely in their advertising departments. If your financial public relations officer frequently dispatches press releases to newspapers and magazines, a TWX machine may be preferable to Telex. But if such releases are only occasional, your Telex machine has the capability of transmitting to TWX through the TCS computer.

## International Wire Communication

Telex is used widely in Canada. The Western Union Telex-TWX directory, issued annually, lists all Telex and TWX users in Canada. The Canadian Telex section lists far more users than the TWX section. Thus, an organization with heavy traffic between the United States and Canadian concerns may choose Telex as the preferred system.

Any Telex machine in Mexico and Alaska can be dialed directly on Telex. No TWX service in available in Mexico or Alaska.

Telex equipment is almost fully compatible with Telex machines used throughout the world. As technology improves, it will soon be possible to directly dial a Telex connection worldwide, just as we now dial a DDD toll call anywhere in the United States. International Telex users abroad utilize a three-row teleprinter which is quite similar to the three-row machine furnished by Western Union to its domestic customers. In contrast, the four-row TWX machine which is used in the United States requires certain operational changes from domestic practice when the teleprinter is used for connection to Telex machines abroad. While the operator can easily learn the minor changes in keyboard character assignments for international TWX transmission, the Telex machine simplifies training of new operators when there is a large quantity of international Telex traffic.

Both Telex and TWX provide quick access to the international carriers for the dispatch of cablegrams (see Chapter 15).

## Conference Calling

In some organizations, it may be desirable to make a simultaneous announcement of a major price change, for instance. TWX offers a conference call service which is set up by the TWX Western Union operator. Telex has no conference facility. However, TCS multiple-address service will dispatch the same text to many different addresses as rapidly as the tapes can be fed into the TCS computer by either Telex or TWX users.

### Accounting for Individual Messages

Up to this point, we have spoken about Telex rates in terms of the rate per minute for Telex transmissions. In actuality, Telex usage is metered for each user on a "pulse" basis. Each pulse is charged at 2½ cents. As the distance increases, the pulses register faster. Thus, in Rate area A, the Telex pulse meter will register seven pulses for a full minute of transmission at 17½ cents per minute. In Rate area D, 24 pulses will be registered for a full minute of transmission at 60 cents per minute. The monthly Telex bill shows the total number of pulses recorded and a cost for the total pulses at 2½ cents per minute. These were the rates in effect in late 1973.

Thus, it usually is not feasible to do direct costing of Telex individual messages, as the Telex invoice does not list the individual transmissions. Those doing such accounting may instead adopt some kind of standard costing procedure based on message length and distance. The message tape is normally used for this purpose. The punched tape contains 10 characters per inch. When the length of the tape is measured with a ruler, an accurate word count can be made which permits the calculation of transmission time.

TWX transmissions are billed on a statement resembling the toll bill which itemizes long distance telephone calls. TWX transmissions simplify cost accounting and provide an exact cost for each connection, but the prospective user must again weigh the accounting advantage against the possible higher cost per message.

### Geographic Location

As mentioned earlier, TWX service is available without mileage charges in almost every location. If Telex service is desired in an area which is not served directly by a Telex exchange, a mileage charge applies monthly from your office to a point 10 miles distant from the city limits of the closest Telex exchange city. The cost of this mileage may offset or exceed the savings in transmission cost otherwise brought about by Telex.

### Western Union 60-Speed TWX Service

At the time Western Union acquired TWX from the Bell System, many TWX units were utilizing three-row keyboards and were operating at 60 words per minute rather than 100 words per minute. While most users of TWX service have now changed to 100-speed machines, there are still many 60-speed machines located throughout the United States. Western Union estimated that by the end of 1973 only sixteen hundred 60-speed teleprinters would still be operating in the United States.

The Western Union Telex-TWX directory gives special instructions on the keyboard modifications which must be made by your operator if she is sending a transmission from a 100-speed machine to a 60-speed machine, and vice versa. Be sure that your operator is familiar with these instructions. The 60-speed equipment is identified in the directory by any area code different from 510, 610, 710, 810, or 910.

Western Union continues to offer 60-speed TWX equipment to its customers. Rental rates differ slightly from those for 100-speed equipment. Transmission rates are billed in full minutes, as in 100-speed service. However, 60-speed equipment does not have access on outward transmissions to the TCS computer, so all the services associated with the computer are unavailable. On balance, Telex would seem to provide a vastly superior service to the TWX 60-speed offering.

## Tel(T)ex Service

If a firm is using Telex and wishes to send a Telex message to another organization, it is customary to first look in the Western Union Telex-TWX directory to determine if the recipient has Telex service. If the recipient is not listed in the Telex section, the dispatching operator will check the TWX section, as a TWX message dispatched via the TCS computer will be the next least expensive alternative. If the recipient does not have TWX service, Western Union offers Tel(T)ex service. The directory lists cities in the United States and Canada offering Tel(T)ex service. The sender dials the number of the Western Union office in the destination city offering Tel(T)ex service. Western Union receives the message on the Telex machine in the distant Western Union office. The message is then relayed by telephone to the recipient. A $3 delivery charge applies to the telephone call under the rates effective in late 1973. The Telex transmission charge between the originating city and the destination city also applies and is included in the usual Telex monthly invoice.

We believe that Tel(T)ex service is an obsolete service offering and should not be used. This is because the telephone service charge is in excess of the cost for a three-minute telephone call to any destination city in the United States or Canada. The sender could easily dispatch a "telephoned telegram" by securing the telephone number of the recipient from the Universal Information operator. The distant telephone will be ringing within seconds. The calling person asks for the secretary of the person to whom the message is addressed. It is read and recorded in longhand. At the end of the message, the recipient is asked to read the message back for accuracy. The name of the recipient, date, and time are recorded on the originator's copy and no discussion on content of the message is permitted. The cost of the call should average far less than the total cost of the Tel(T)ex message. If the message is not urgent enough to require delivery during the same day, Mailgram service can be used if it is less expensive than the telephoned telegram.

Tel(T)ex service is *not* available to TWX customers.

## MAILGRAM SERVICE

The Mailgram is a joint venture of Western Union and the United States Postal Service. It is designed to provide next business day delivery to any point in the contiguous 48 states and to Alaska. This delivery standard is being met through excellent teamwork on the part of both venturers. Here are some Mailgram features.

### Cost

A TCS computer charge is calculated on the distance between your office and a destination post office which has been designated as a receiving point for Mailgrams addressed to your recipient's zip code. From Telex machines, this charge varied in 1973 from 1¾ cents per $\frac{1}{10}$ minute in Telex Rate area A to 6 cents per $\frac{1}{10}$ minute in Telex Rate area D.

From TWX machines, the charge varies from \$0.0265 per $\frac{1}{10}$ minute in area A to 9 cents per $\frac{1}{10}$ minute in area D. This rate differential is based on the difference in transmission speed between TWX and Telex. At 100 words per minute, your TWX machine will be connected to the computer for less time than would be your Telex machine at 66.67 words per minute.

In addition to the charges per $\frac{1}{10}$ minute, a flat delivery charge of 75 cents applied in 1973.

### Mechanics of Transmission and Delivery

The Mailgram may originate on a Telex or TWX machine and must be routed via the TCS computer. Users of Info-Com service, which is described later in this chapter, can likewise originate Mailgrams. A special format is used by the operator. The format must include the zip code of the addressee. The Mailgram must be dispatched by 7 P.M. *destination time* for next-day delivery. This means that Eastern time zone originators must dispatch Mailgrams to the Pacific time zone prior to 10 P.M. Eastern Standard Time.

The receiving post office tears off the Mailgram from a high-speed printer after the printer writes the message on a telegraph-type form. It is placed in a distinctive envelope and given speedy attention so that it goes directly to the postal carrier who will make the first daily delivery to the office or residence of the recipient. In some cases where Mailgrams are dispatched early in the business day, delivery may be made the same day if the message is routed to an area which has more than one daily delivery.

### Voice-originated Mailgrams

Western Union has established an "800" inward WATS service which can be used by the public for dispatching Mailgrams, telegrams, night letters, and cablegrams. The Mailgram rate for 100 words or less in 1973 was $2. For each additional 100 words or fraction thereof the charge was $1. Mailgrams are also accepted over the counter at Western Union offices and can also be sent over local tie lines to Western Union offices. If the person telephoning the Mailgram wishes a confirmation copy mailed to him, a 50-cent fee is charged for this special service. A Mailgram confirmation costs 75 cents.

### Computer-originated Mailgrams

A minimum charge of $200 was applicable in 1973 for transmission of multiple Mailgrams from punched cards or magnetic tape. The first 600 characters, or approximately 100 words, are billed at 80 cents. Each additional 600 characters is billed at 25 cents. Thus, the $200 minimum charge is equivalent to about 250 average-length messages. Tape acceptance centers are located in major cities and Western Union is expanding the list of cities in which this service is available.

### Computer Acceptance of Multiple Messages

The TCS computer will accept up to 50 messages on one connection. The messages may have different addresses and different texts. Or, you may transmit a single text and up to 50 different addresses. Thus, the Mailgram can be an excellent substitute for the expensive "book" messages formerly sent as telegrams or night letters.

### Contrasting Mailgram with the "Telephoned Telegram"

In the section on Tel(T)ex I described the use of the telephoned telegram, a message which is telephoned to your recipient over toll or WATS facilities by a member of your clerical staff. The Mailgram is a written message service. A typical cost for a Mailgram of $1, for instance, is equivalent to an interstate toll call which travels

430 airline miles. Consequently, you must ask if the written nature of the Mailgram justifies a cost that may be higher than the telephoned telegram which is delivered by telephone immediately rather than the following business day. You may choose to furnish your message center or message originators with a map on which a circle is drawn around your originating location to indicate the area that can be reached less expensively by the telephoned telegram. Then your originators can decide if they require that a written message be delivered at the distant point or whether a telephone call would do the same job at less cost. It is very important to have such telephoning done by a member of the clerical staff, *not* the message originator. This is because the originator will tend to engage in social conversation with his recipient at a considerable waste of time and money. If the conversation extends to the usual United States average of six to seven minutes, the Mailgram is a real bargain regardless of distance.

If you are about to acquire Telex or TWX service chiefly for the use of Mailgram service, remember that accurate costing requires that you prorate your equipment rental cost over the cost of the messages you dispatch.

### Contrast of Mailgram to Telegrams and Night Letters

As a 100-word night letter was billed in 1973 at $3 for telephoned delivery and $6 for physical delivery, the Mailgram is obviously less expensive and is equally effective as a distinctive written message.

In 1973,Western Union's physical delivery standard for a full-rate telegram was five hours. The cost for a 15-word physically delivered full-rate telegram was $6.75. If it is not necessary for your message to be delivered within the five-hour standard, then the Mailgram will provide an excellent cost reduction.

### Maximum Length of Mailgrams

The Western Union computer will accept a Mailgram which contains up to 15,000 characters, or approximately 2,500 words.

### Certified Mailgrams

By following instructions furnished by Western Union, the Mailgram user may direct the delivering post office to handle the Mailgram as certified mail. If desired, a return receipt is furnished showing to whom the Mailgram was delivered and the date of delivery. If the recipient is not available when the postal carrier calls and if the sender has specified "delivery to addressee only," the

Mailgram is returned to the postal delivery office, where it can be claimed by the addressee. If the addressee has moved, leaving a forwarding address, the post office will forward the Mailgram.

### Undeliverable Mailgrams

The post office will hold an undelivered certified Mailgram for seven days. Then it is returned to the Mailgram Service Center at Middleton, Virginia, which sends it to the sender. Noncertified Mailgrams which are undeliverable are returned immediately to the Mailgram Service Center for return to the sender.

## WESTERN UNION PUBLIC MESSAGE SERVICE (PMS)

For many years, the full-rate telegram and the telegraphic night letter were Western Union's most profitable services. In earlier times, the person sending a telegram could expect delivery in less than an hour. Overnight messages were delivered promptly in the early business hours of the following day. Western Union equipped volume users with teleprinters connected to the local Western Union office or with "Desk-Fax" facsimile machines. There was no charge for such equipment.

In late 1973, Western Union was still advertising its telegram service. However, in a few localities where teleprinters and Desk-Fax units were still installed, a monthly rental charge applied to such equipment. Service standards were lowered substantially. In late 1973, Western Union guaranteed delivery of full-rate telegrams within three hours by telephone or tie line and within five hours by messenger.

What is the cost of telegraph service? Typical costs for sending telegrams and night letters via the TCS Telex/TWX computer have been shown in Table 1. But organizations without Telex or TWX service may still wish to send a telegram or night letter. In late 1973, the following rates applied to interstate messages (intrastate rates may be slightly different):

Full-rate telegram ("day telegram"): first 15 words, $3.95; each word from 16 to 234 words, 12 cents; each word over 234 words, 8 cents.

Night telegram ("overnight telegram" or "night letter"): first 100 words, $3.20; each word over 100 words, 3 cents.

For physical delivery within the normal delivery area of the destination city: $3.

For physical delivery beyond the normal delivery area: messenger or taxi fees as required.

For physical delivery to destinations in Canada: full-rate telegram, $3; night telegram, $1.50.

For mailing a confirming copy to the person telephoning the telegram to Western Union's "800" inward WATS service: full-rate or night telegram, 50 cents. For confirmation by Mailgram: 50 cents.

For *collect* messages: a surcharge of 50 cents billed to the *sender*.

Delivery charges do not apply to "personal opinion" messages offered at a special rate or to messages which can be routed to the recipient over a tie line connected to a receiving teleprinter or Desk-Fax facsimile printer.

To ensure physical delivery, the words "deliver only" should be incorporated in the first line of the address; otherwise Western Union will choose between its choices of delivery methods. Messages of condolence, congratulation, casualty, or similar content will usually be delivered at the additional charge.

Despite the cost and relatively slow delivery speed of the traditional Western Union day and overnight telegrams, these messages are still used by some organizations which have not yet analyzed the possibilities of using alternative services. In fact, Western Union reported that in the first two quarters of 1973 telegram volume increased over previous quarters.

The administrator of telecommunications services should ask his accounting department for a copy of recent billings from Western Union. Then he should identify those who are sending telegrams. For instance, many sales managers have used "book" telegraph messages to announce price changes. A single text is composed. Western Union's local office is given a list of dealers, agents, and salesmen. The same message is sent to the entire list. Frequently the originating executive does not specify overnight telegram service, so the messages travel as full-rate day telegrams. At a delivered cost of $6.75 for a 15-word message, the announcement may result in an invoice from Western Union for hundreds or even thousands of dollars.

The administrator can interview the telegram originators. He can point out the economies of Telex, TWX, and Mailgram. He can explain the speed and economical nature of the telephoned telegram described earlier in the Tel(T)ex section of this chapter. He then encourages his originators to continue to write brief messages but to send them to the company's message center. There a trained supervisor will scan the message, edit it where necessary, and see that it is dispatched by the least expensive method consistent with the nature of the message and the delivery date and time specified by the originator on his message form. A carefully designed form will help guide your message originators by requiring an indication of acceptable delivery by telephone or by physical delivery today or tomorrow. If the message center operator uses the telephoned telegram technique, it may be desirable to mail a confirmation copy to the addressee, just as Western Union mails a confirmation copy after telephoning the text to the addressee.

### Required Uses of Western Union Telegrams

In past years, certain contractual documents have specified the use of the Western Union telegram. A telegram may seem to be required in bidding on contracts or in confirming participation in an investment syndicate. Some accounting or legal matters may best be handled by a message in telegraphic form, written on a telegraphic-type message blank and physically delivered by some independent means.

If such situations exist in your organization, try to determine whether an existing contractual requirement can be modified to permit the use of the Mailgram or a Telex/TWX message. In matters involving termination, layoff, or recall of employees, certified mail with return receipt requested provides a proven delivery record, which may satisfy union contract requirements.

## WESTERN UNION TELEPRINTER EQUIPMENT

In the preceding sections I have discussed the cost of various types of transmission on Western Union's array of network services. As a transmitting device is necessary, you will want to determine whether you would like Western Union to furnish a teleprinter machine or whether you would like to use a compatible machine already available in your office. Another alternative is to purchase or rent a compatible machine from an outside vendor.

Western Union has recently separated its teleprinter rental charge into two parts. One part covers the cost of "network access," and the other part covers the rental cost of the machine you select. Outside vendors will furnish equipment which can be used on the Telex or TWX network. Also, you may use a four-row teleprinter furnished by a shared-time computer organization for access to TWX. If usage is to be heavy, you should ask outside vendors for quotations on heavy-duty equipment. Here is a brief description of equipment characteristics available from either Western Union or outside vendors.

*KSR teleprinter:* This unit is used "on-line." The operator must dial the connection, then use the keyboard to manually type the message after the two Telex or TWX machines exchange their answer-backs to ensure that a correct connection has been established. While the KSR unit is connected to a circuit which will transmit at 66.67 or 100 words per minute, manual typing by the operator will probably reduce this speed to approximately 30 words per minute. Consequently, KSR equipment can rarely be justified. This is because the slight saving in rental or acquisition cost is quickly offset by increased transmission cost from the slow manual typing operation.

*ASR teleprinter:* This unit is similar to the KSR, but a tape punch and a tape reader are added. The tape punch permits the operator to prepare a tape on a "local" circuit without dialing the connection. As the operator types on the keyboard, the message appears on the platen, so it can be checked for accuracy. If a mistake is made, the tape can be corrected. When the tape is complete and correct, it is torn off and inserted in the tape reader, the connection is dialed, and the machine transmits at its full speed of 66.67 or 100 words per minute.

*RO teleprinter:* This unit is used for "receiving only." Thus, no keyboard is necessary.

The basic units I have described can be modified to accomplish special purposes. For instance, speed can be adjusted to 60, 66.67, 76.67, or 100 words per minute. It is possible to supply a typing reperforator which stands alone as a separate unit. As messages are received, a five-level or eight-level tape is produced which can then be used in data processing activities, or such equipment can be incorporated in the teletype unit itself. The marketing representatives of Western Union or the independent vendors can give you complete details.

The Western Union machines come in certain model designations. It will help you understand Western Union technical terminology if I explain the model differences:

*Model 28:* Heavy-duty units operating at speeds of 60, 66.67, 76.67, or 100 words per minute and using the five-level Baudot code.

*Model 32:* Standard-duty units usually operating in Telex service at 66.67 words per minute, but usable in other services at up to 100 words per minute. Model 32 uses the five-level Baudot code. See Figure 1.

*Model 33:* Standard-duty units operating at up to 100 words per minute, using the seven-level ASCII code with the option of adding the eighth level if "parity" is desired in a data processing application. See Figure 2.

*Model 35:* Heavy-duty units operating at up to 100 words per minute, using the seven-level ASCII code with the option of adding the eighth level, as in Model 33.

*Model 37:* Heavy-duty private line service using the ASCII code. These machines can operate on private networks at up to 150 words per minute. They can offer special features for tabulation and other functions helpful in order or data processing operations. Model 37 prints upper and lower case alphabets plus 42 punctuation marks and symbols, a total of 94 printing characters.

These machines are used in both Telex and TWX operations. Also, as I will describe later, Western Union offers private line teletype service on a point-to-point or network basis, and this equipment can be used for such service.

Supplementary teleprinter equipment is likewise available from Western Union. Some of the commonly used items are:

Computer interfaces for Telex, TWX, and broadband exchange services
Alternate-use adapters for using TWX teleprinters on long distance voice circuits
Automatic dialers
Transmitter control arrangements (see next section)
Sprocket feeds
Form accumulation shelves

Rates for Telex and TWX equipment are best procured from your local Western Union marketing representative. In late 1973, access to the network was billed to Telex customers at $24 monthly and to TWX customers at $11.50 monthly. A $50 installation charge applied.

The Telex teleprinter monthly rental rates varied from $34.50 for a Model 32 KSR to $86 for a Model 28 ASR. TWX equipment rates were higher, ranging from $47 for a Model 33 KSR to $123.50 for a Model 35 ASR. Installation charges for all units were set at $50.

## SOME SUGGESTIONS FOR OPERATING TELEX AND TWX EQUIPMENT

I consider Telex and TWX to be splendid services which can be of great usefulness to the user of telecommunications services. Here are some suggestions which will help you employ Telex and TWX to best advantage.

### Calculate Cost per Message

If you install one or both of these services, remember that the equipment cost is quite substantial. Do not be misled by low cost per transmission, as you must add the prorated cost of machine rental to your cost per message to secure a valid comparison between the cost of Telex-TWX and that of a telephone call or an ordinary telegram or night letter. The more you use your equipment, the lower will be your unit cost. On the other hand, one of your department heads may decide that he needs TWX for his press releases to financial publications. If he sends out only a few releases monthly, you may find that you will be paying a very high cost per message if you prorate the equipment rental cost over the messages you send. If you equip your distant offices with Telex or TWX to help them send messages to headquarters, the rental cost on those machines must likewise be considered. In situations involving light usage, you may wish to use the telephoned telegram technique mentioned in this chapter's section on Tel(T)ex.

### Prepare Tape on All Transmissions

You should constantly monitor your dispatching operators to make sure that tape is prepared "off-line" on even the shortest messages. Tape is dispatched automatically at 66.67 words per minute in Telex service and 100 words per minute in TWX service. Even a competent teleprinter operator rarely exceeds a 30 word per minute typing speed. Your billing for transmission cost could double if you permit manual typing on a connected circuit to the destination machine.

### Control Machine-to-Machine Conversation

Once a direct connection is established, either Telex or TWX permits "conversation" between the two machines. In some organizations, an executive will stand behind an operator and dictate replies to questions being asked at the other end of the circuit. If lengthy discussion is needed, it may be better to make a DDD toll call or WATS call, then confirm the conclusions of the discussion in the written message. The manual typing operation will increase cost markedly.

### Examine "Header" Format

Basically, you need only to exchange answer-backs to establish the fact that you have dialed a correct number. All Telex machines and every TWX machine that operates at 100 words per minute have a mechanism operated by a teleprinter key which will print the identity of the subscriber at the beginning of the message. The identity is established by the assigned answer-back. For instance, *SUNCHEM NYK* would be an answer-back code which might identify the executive offices of the Sun Chemical Corporation in New York City. The answer-back code is limited to 20 characters, and it is almost always in letters rather than numerals. In TWX operation, when a connection is established with a distant machine, the distant answer-back will automatically appear on the platen of your machine, confirming that you have reached the right destination. In Telex operation, the operator depresses a key which will register your answer-back on the platens of both machines. She then usually depresses two keys in sequence which will cause the distant machine to print its answer-back on both platens. Answer-back codes are made to order for each user of TWX or Telex service.

Many firms add other material to the "header" before starting to print the actual text of the message. Sometimes a complete name and address are given. The word "attention" may precede an executive's name. The date and time of day may be shown. None of these data are required by Western Union's technology. The date may be helpful in filing message copies after transmission, but such a date could be rubber-stamped or inserted by an automatic dating or numbering machine. A message number can be used for accounting purposes. But there is no need for taking circuit time for printing a corporate name and address when the automatic answer-back provides the needed identification. Chargeable time in Telex and TWX starts as soon as the called machine answers. The administrator should carefully examine the beginning of each Telex transmission and instruct his operators on the correct "header" technique.

## Consider Supplementary Equipment for Unattended Service

Both Telex and 100-speed TWX equipment operate unattended. In other words, no operator needs to be present at the distant end, as the receiving machine will automatically receive the message and it will be disconnected when the sending machine disconnects itself from the circuit.

However, an operator must be present at the sending machine. She must wait until the prepared tape has been transmitted. Then she manually disconnects her machine. To conserve the time of the operator, Western Union can furnish a supplementary low-cost device which will disconnect the machine from the circuit when transmission has been completed. In effect, this provides two-way unattended operation. Your operator must be present only to dial the connection and to make sure that the tape has started through the tape reader.

Another low-cost device is available which will permit the operator at the distant end to load a tape into the machine so that it will be ready for dispatch when headquarters calls that particular office. No operator need be present at the distant point. Such devices can be used for after-hour transmission of data or orders. However, it is important that the distant machine be given an unlisted number to prevent outsiders from dialing the listed number of the machine and thus receiving confidential data.

### Keep Machine Available for Incoming Traffic

If you are using more than one Telex or TWX machine, do your tape preparation on the secondary machine or rent a tape perforator as a supplementary piece of equipment. Thus, you keep your listed Telex or TWX number available for incoming calls and avoid giving your callers busy signals.

### Consider "Blind" Tape Preparation

All authorities on tape preparation agree that an operator can prepare tapes more accurately if he is not distracted by reading what he is writing on the platen of the machine. Speed and accuracy are increased by "blind" operation. In addition, the platen of the machine is available for recording incoming traffic. In normal Telex operation, an incoming call to your Telex machine will interrupt your preparation of tape for an outgoing call. If you would like to cut tape "blind" and thus free your machine for important incoming traffic, you can substitute a more expensive teleprinter which will give you this advantage. Similar arrangements can be made for TWX transmissions.

### Consider Elimination of Final Answer-backs

In training operators for its customers, Western Union emphasizes the desirability of exchanging answer-backs at the end of the message. The operator presses a key which will identify the calling machine and the called machine will also respond with its answer-back. This exchange indicates that the transmission has been satisfactorily received. As an answer-back phrase may contain up to 21 characters, it is equivalent to more than three words. A reduction in chargeable Telex or TWX time may come about through elimination of the final answer-back exchange. Instead, your operator can press the "stop" key and the connection will be terminated. Western Union feels that the final exchange of answer-backs verifies that the message has been received and that the additional pulse charge of approximately 2½ cents is justified by your receipt of this assurance.

### Consider Simplification of Signature

Any words appearing in the signature of a Telex or TWX message are counted as part of the chargeable time of the transmission. Your organization is identified automatically by your assigned answer-back, which appears automatically at the beginning of the message. You should be on the alert for lengthy titles, department names, or addresses which originators may append to messages and which untrained operators may incorporate into the tape.

### Eliminate Unnecessary Operation

As the Telex pulse meter or the TWX clock is operating throughout your transmission, your operator should avoid using chargeable time for nonproductive operations. These include:

*Tabulation:* Some department heads may insist on elaborate column tabulation of statistical material. This takes excessive machine time. Perhaps you can convince those men that the tabular material should be put in paragraph form and tabulated at the receiving end. Or perhaps the material can be sent by facsimile or airmail transmission at substantial savings over tape-cutting labor and Telex or TWX transmission.

*Punctuation and numerals:* As Telex operates on a three-row keyboard, any numbers or punctuation require a carriage shift which takes chargeable time. The figure "2" is a single stroke, but two carriage shifts are required, so the word "two" would result in the same chargeable time. Excessive punctuation should be eliminated if not necessary for message clarity.

### Eliminate Spaces

All typists are trained to add a space after a period is struck. Where possible, eliminate spaces, as each manipulation of the teleprinter space bar increases the transmission cost.

### Use the Full Writing Line

When the operator comes to the end of a sentence and is two-thirds of the way across the platen, she may be tempted to operate the line space key and start the next sentence on a new line. The line-spacing operation takes time and increases costs, especially in Telex transmission billed in $1/10$-minute increments.

### Develop Key-Stroke Conservation Techniques

A judicious use of commonly understood abbreviations may cut transmission cost markedly. For example, "week" becomes "wk" and "statement" becomes "stmt." You should read your own message traffic and make a list of frequently used words which can be abbreviated by both your message originators and your dispatching operators.

### Consider Editing Procedure

We suggest that you appoint a qualified person as supervisor of the message center. That individual reads each message reaching the department. If the message is unclear, the originator is telephoned for clarification. The supervisor has the authority to delete unnecessary words—adjectives, prepositions, connectives. If simple rewriting can decrease cost without changing the sense of the message, she is authorized to do that. If the originator specifies a dispatching method which is needlessly expensive, the supervisor will use the correct method or will clear her suggested method with the originator or his department head. We have found that such a trained person regularly saves several times her salary in reduced transmission costs.

### Consider Training in Concise Writing

Just as a trained supervisor saves money for her employer, a trained message originator will save even more. Here is a procedure that telecommunications administrators can employ to show people how to shorten their wire messages. Select messages which are wastefully composed. Eliminate the name of the originator, to avoid embarrassment, and then duplicate the message and distribute it to company personnel who wish to improve their skill in concise writing. After the group has been given half an hour of instruction, which is based on the domestic and international word-count rules for telegraphic messages, they rewrite the messages. In almost every case you will find that people can express the same thought clearly in 50 percent less words. Further, they remember what they have learned, so you need only train new personnel. Obviously, the 50 percent word-count reduction has a substantial impact on the organization's transmission costs.

### Locate Wire Dispatching Equipment in One Place

Many department heads, learning about the usefulness of Telex and TWX equipment, will want a machine of their own in their department. Each such installation results in additional equipment rental costs. Equally important, it is more difficult to train many different machine operators than to concentrate training and traffic on a few machines in one location. I think it is much better to provide speedy access to the message center (see next section) and then convince executives about the cost and service advantages of the centralized message center.

### Provide Adequate Messenger Service

Organizations may spend thousands of dollars on the transmission of urgent wire messages and then let those messages languish in an "out box" for hours. You should check operations in your message center and make sure that urgent incoming messages are expedited for immediate dispatch. The dispatch may be by telephone to the affected individual or department. If time does not permit reading and recording of the text, a messenger from the department can visit the message center to pick up the message. Hourly message center messenger pickup should be required for nonurgent messages.

Conversely, prompt pickup and delivery should be assured for messages to be dispatched from the message center. If messenger service or personal delivery is not feasible, some type of telephone recording device or facsimile service can be very helpful. In a number of companies, a message originator can dial a three- or four-digit number to connect him with recording equipment used only for wire messages. The operators in the message center watch the equipment closely. The belts or disks are removed promptly; a rough draft of the message is transcribed. The supervisor scans the message for clarity and economy and marks corrections; a tape is then prepared on the dispatching equipment.

### Resist Multiple Copies

Western Union supplies teleprinter rolls which provide defined numbers of copies. In some companies, the originator's message draft is returned to him with a time stamp indicating the date and time of dispatch. In other companies, the originator receives the copy made when the tape is prepared; the message center keeps the second copy, which is prepared as the tape is dispatched. In such cases, the originator's input copy is filed with the copy of the transmitted message. This type of handling on outgoing messages eliminates the need for multiple-copy teleprinter paper.

On incoming message traffic, an original and one copy may be sufficient. Any request for multiple copies should be examined carefully. Under most circumstances, I recommend that the person receiving the message should arrange for any multiple copies which are required. Alternatively, an inexpensive copying machine in the message center will produce copies at less expense than buying multiple-copy paper rolls. An alert telecommunications coordinator will occasionally make a thorough survey of wastebasket contents in the message center to see what is being thrown away.

## ADDITIONAL WESTERN UNION SERVICES

In the preceding pages of this chapter, I have described the Western Union services which have the widest application to those who read this book. In this portion, I will discuss other Western Union services which may not be as well known and which may prove very useful to organizations with special requirements.

### Hot Line Service

Hot Line is a point-to-point voice communication service. It is offered only in certain cities. It operates on a 24-hour, 7-day basis. A telephone handset is installed on the desks of those wishing to use the line. When the transmitter is picked up, the telephone on the desk of the distant person rings. No dialing is required. Connection is almost instantaneous. The service is billed in $\frac{1}{10}$-minute increments, plus a monthly rental charge for access to the channel and for the instruments involved.

We believe that Hot Line is an excellent service which may save a considerable amount of money over the disbursement which would be required for a full-time voice tie line between the cities involved. In 1973, Hot Line could not be terminated on a Bell System switchboard, but certain independent companies permitted such termination. Where switchboard termination is not permitted, Hot Line serves individuals at specific desks. Up to five extension instruments can be used with the primary instrument, although Western Union recommends that no more than three instruments be installed. The extensions permit other individuals to use the Hot Line circuit.

Hot Line will undoubtedly face competition from FCC-regulated common carriers who will offer similar service over microwave circuits. Thus, the prospective user should check the competition carefully. Here are a few additional facts and statistics. The rates were those in effect in 1973.

*Cost:* The basic service charge is $50 monthly at each end, including the first instrument. For the first extension instrument it is $3.50 monthly. Additional charges are made for multibutton instruments, lamping, and intercom. Transmission charges vary from 15 cents per minute (example: New York to Stamford) to 65 cents per minute (example: New York to San Francisco or Los Angeles). The usage time is billed in $\frac{1}{10}$ minutes. No minimum charge applies.

*Executive exclusion:* Confidential conversations cannot be heard on extensions.

*Telco interconnection:* If desired, the normal telephone instrument can be used through interconnection arrangements.

*Extension line service:* The service can be extended to suburban offices.

*Camp-on:* Western Union advertises that connections will be made immediately in 90 percent of attempts. If connection is not made, an automatic camp-on will signal the callers as soon as the circuit is available.

*Hot Line termination:* Hot Line does not require a separate telephone instrument as it can be terminated on a telco button-equipped telephone if desired.

### Broadband Exchange Service (BEX)

This service provides for furnishing a four-wire circuit which can be used for voice, data, and facsimile transmission. Voice usage can be coordinated with data transmission through appropriate interface devices. The service is billed in $\frac{1}{10}$-minute increments after the first minute of transmission. Direct connections had been established in 1973 to 49 United States cities, and arrangements can be made to extend the service to other cities by using leased circuits furnished by Western Union. Canadian transmissions can also be sent from the United States. Two types of service are offered, the first from a speed of 50 to 600 bauds, and the second, from 10 to 2,400 bauds. The receiver can signal the sender at any time. Push-button dialing is furnished as standard equipment. Full-duplex service is optional. The system can operate on an unattended basis. A direct connection can be established to international carriers furnishing Datel service for data (see Chapter 15).

Your data communications experts may find BEX to be of great interest because it is noncontractual in nature and is billed on a usage basis plus a nominal charge for rental of the equipment involved. The data sets used may be furnished by the user or by Western Union. Any digital apparatus can be used on BEX, which permits the transmission of punched cards, paper tape, magnetic tape, computer connections, and facsimile. In 1973, the cost per minute ranged from 15 to 75 cents, depending on distance and which of the two types of service was utilized. Western Union claims that BEX offers a fast turnaround time for those who use a polling technique to secure data from distant points. A volume discount is offered which may make BEX less expensive than the use of toll or WATS circuits for this type of data transmission.

### Datacom Service

This service was designed by Western Union to fill a need for a low-cost bulk transmission service for data. It is available to and from major United States cities. The service was offered in 57 cities in 1973, and it can be extended to other cities by utilizing leased lines. It is furnished on a 24-hour, 7-day basis with a one-month minimum service contract. Half- or full-duplex service is offered at various speeds. Western Union manages all equipment and circuits and guarantees the quality of circuits utilized. Customer-provided equipment is permitted.

If your data people are interested in acquiring private line facilities which will be used exclusively for data, you should suggest that they investigate Datacom as an alternative to the usual conditioned or normal voice grade circuit rented from the telephone utilities. As time goes on, FCC-regulated common carriers will be competing with Western Union for such business, so those carriers should also be contacted for specifications and quotations.

### Info-Com Service

Western Union has developed a national system of interconnected computer-controlled message switching centers. Info-Com utilizes this system. Western Union engineers will plan a private network system for the subscriber which is dedicated to 24-hour, 7-day use. The private network uses the national system on a shared-time basis.

Any organization which has sufficient traffic on telegraph-type circuits to justify an investigation of private line facilities should consider Info-Com as an alternative to acquiring a private line telegraph-grade network from either Western Union or the telephone utilities. Three classes of service are provided, depending on the volume of transmissions. Usage charges are based on a minimum charge plus overtime charges at reduced rates when the subscriber uses Class 1 or Class 2 Info-Com service. Class 3 service utilizes the domestic Telex network and is billed on a per-minute basis. A monthly charge is made for the basic system.

Western Union offers this list of Info-Com advantages:

*Interservice capability:* Info-Com access to Mailgram, Telex, TWX, and PMS.

*Canadian service:* Customers' plants and offices in Canada can communicate with the United States.

*Priority:* Special "precedence" service is available for rush messages.

*Message retrieval:* Up to 60 days' retrieval is available.

*Status and traffic load reporting:* Traffic contributed by each network station is available, plus general traffic load and distribution reports. Thus, the makeup of the network configuration can be evaluated regularly.

*Multiple addressing:* Info-Com will accept a single text for multiple addresses.

*Alternate delivery:* If a particular station is out of service, the message can be delivered to an alternate station.

*Alternate routing:* Western Union supplies alternate channels between computer centers to avoid delays due to circuit outage.

*Choice of codes:* Class 3 service uses the five-level Baudot code; Class 1 or Class 2 service uses the seven-level ASCII code.

*Privacy:* Western Union assures the customer that no other user of the shared system can secure access to his data.

*Itemized billing:* The customer is billed for equipment and usage cost by each terminal, so costs may be allocated if desired.

We believe that Info-Com appeals particularly to the organization which has widely varying quantities of data for transmission to or from a variety of stations. When a dedicated network is acquired, the charge for the system is independent of the amount of data transmitted. Info-Com permits the user to control expense by renting a class of service based on the quantity of data transmitted from a specific station.

## Sicom Service

This service is designed for the securities industry. It transmits large volumes of short messages through the use of a computer-controlled message switching system. Standard order formats of the New York Stock Exchange are used. Western Union can give full details to any prospective user.

## Western Union Leased Services

Western Union competes with the telephone utilities and with the newly emerging specialized common carriers in furnishing a variety of private line services. The service may be a simple point-to-point transmission or a complicated network. Both telegraph-grade and voice-grade services are supplied. Switching systems can be furnished which automatically route a message to its destination through typing a code symbol. Polling can be arranged to determine if a particular station has traffic ready to be sent. Western Union classifies its leased transmission services in series packages:

Series 1000—Teleprinter
2000—Voice
3000—Data
4000—Facsimile
5000—Telpak
8000—Wideband

Circuit conditioning and alternate use arrangements are available as required.

Any organization which requires its own private voice or wire network should ask Western Union to study the problem and submit a quotation. The Long Lines department of AT&T will also wish to quote through your telco marketing representative. FCC-regulated common carriers will also quote if your geographic pattern meets the requirements of their system. In general, the cost of the system is based on the airline miles between points plus the charges for service terminals and equipment. Customer-provided equipment may be used.

### Miscellaneous Services

**Dealer Inquiry Service.** A 24-hour, 7-day service provides a list of local offices or dealers to prospective customers who dial a Western Union toll-free "800" number and ask for "Operator 25."

**For Your Information (FYI).** This service is available to Telex and TWX subscribers. Four news services are provided: News, Finance, Market, Sports. The user dials a special number, "1213," and receives a summary report on that connection. The charge for the service is the normal pulse charge on the connection. News summaries are updated hourly. On request, Western Union will arrange to send the requested summaries at a prearranged time between 9 A.M. and 8 P.M. The charge for this service is based on the TCS collect computer rate schedule. Reuters News Agency prepares the news dispatches.

**Candy-Gram.** This service arranges for the acceptance and delivery of specially boxed candy accompanied by a telegram. Deliveries are made from local Western Union offices by messenger or by mail. Orders must be given to a local office.

**Dolly-Gram.** This service provides for the delivery of one of two types of dolls. A "Happy Birthday" doll or a "Just To Say" doll is delivered at a fixed price. A choice of 22 greeting cards is available with the "Just To Say" doll. A telegraphic notice of delivery is sent to the sender at extra cost.

## INTERCHANGE OF MESSAGE TRAFFIC BETWEEN TELEX, TWX, AND FACSIMILE

There has been an increasing use of facsimile transmitting equipment to transmit typed, written, or drawn copy between facsimile users (see Chapter 12). As vendors and lessors of facsimile equipment have cooperated with users in furnishing directories of users, it occurred to at least one entrepreneur that a facsimile-Telex-TWX interchange service would be helpful to those concerned. This service became operational in late 1973. It was named "T-FAX" and was sponsored by a private organization.[1] That vendor advertised these service features:

1. Provides communication between a Telex or TWX terminal and a facsimile terminal, and vice versa.
2. Provides exchange of alphanumeric information, but does not have graphics capability.
3. Provides a delivery standard of two hours.
4. Eliminates duplication of terminals. If a facsimile machine is not needed for an organization's own use, but that organization wishes to send messages to the facsimile terminals of other organizations, T-FAX provides the needed service. In effect, the service extends the number of terminals which can be reached by TWX, Telex, and facsimile users.

The rate structure applicable in 1973 was:

From Telex or TWX to facsimile: 1½ cents per word, with a $1.50 minimum charge (100 words)

From facsimile to Telex-TWX: 2 cents per word, with a $2 minimum charge (100 words)

Monthly maintenance charge for billing number: $10

The service operates as follows: Each subscriber to the service is given a "user number" which is transmitted as a part of each message sent to the T-FAX terminals. A facsimile user calls an inward WATS "800" number which connects him to the T-FAX terminal. The Telex or TWX subscriber dials an appropriate Telex or TWX number to connect with the T-FAX terminal. At the T-FAX terminal, automated equipment checks the correctness of the billing number. If the incoming message consists of facsimile copy, a tape is prepared and dispatched to the indicated recipient on Telex or TWX equipment. If the incoming message appears on the platen of Telex or TWX equipment, it is removed from the machine, placed in a facsimile transmitter, and forwarded to the indicated addressee.

[1] Information Management Inc., 5700 Broadmoor, Mission, Kans. 66202.

# CHAPTER FIFTEEN

# INTERNATIONAL TELECOMMUNICATIONS

Do you transact business outside the North American continent? Do you send an occasional cablegram to friends overseas? Do you make international telephone calls? A book on telecommunications is not complete in a world of shrinking distances unless the reader is given some guidance on the economics and complexities of international communication. Such communication has been speeded and improved constantly by the laying of new cables and the increased utilization of satellite circuits. Rates have steadily decreased—a remarkable achievement when we view recent increases in charges for domestic long distance telephone service. This chapter is designed to help you reduce cost and improve service on overseas voice and written message requirements.

Many different methods of communication are available to international points. In Chapter 13 I discussed one of the least expensive methods: airmail and the air-letter sheet. In this chapter I will describe other methods, including Message Telecommunications Service (MTS), international Telex, international telegrams (cablegrams), teleprinter leased channels, voice/record leased channels, Hot Line, Datel, photo-transmission, closed-circuit television, and marine telegrams.

Canada is an international destination from the United States, but it is excluded from this chapter because it can be reached by the domestic long distance network and by domestic Telex and TWX service, as well as by voice or data private lines and FX service. Alaska and Mexico can be reached by the long distance network and Telex, but not by TWX. Western Union domestic telegrams and night letters can be used to points in Canada, St. Pierre and Miquelon, Mexico, and Alaska. The international carriers likewise provide service from the United States to Mexico, Alaska, and St. Pierre and Miquelon, but rates and word-count rules follow the domestic telegraph pattern rather than the international pattern. The Philippines, Hawaii, Puerto Rico, and points in Central and South America are reached through the international communications carriers. Some special word-count rules apply on telegram traffic to Mexico, Canada, and St. Pierre and Miquelon. There is a 10-word minimum, rather than the domestic 15-word minimum. Any arabic numeral is counted as one word in a telegram to Canada or St. Pierre and Miquelon. Thus, any message containing many figures can be transmitted much less expensively by Telex or by the "telephoned telegram" technique described in Chapter 14.

## MESSAGE TELECOMMUNICATIONS SERVICE (MTS)

The Long Lines department of AT&T manages the sending of telephone calls to international destinations. Those calls are being connected in less time as transmission technology changes and improves. Also, rates have been reduced. Thus, the busy executive may be tempted to pick up the telephone to discuss a matter which formerly would have been covered in an exchange of cables or international Telex messages.

In Chapter 14 I discussed the subject of converting the telephone message to a written message, which costs less to transmit. The same principles apply to international telephone service. At rates in effect during 1973, the least-cost international Telex message was almost always less expensive than the least-cost international telephone call.

What about the quantity of words which can be exchanged in relation to cost? In the three-minute basic telephone rate period, two persons speaking normally can exchange a total of 500 words —and more if they are rapid speakers. If an executive is dictating, he must adjust the speed of talking to his shorthand stenographer's ability, which seldom exceeds 130 words per minute, or 390 words in three minutes. A recording machine may take faster dictation, but, in fact, the careful dictator must slow down to spell proper names and repeat figures. These factors must be considered when an international telephone call is contrasted to the efficiency of international Telex operating at 66.67 words per minute, or 200 words in three minutes. Perhaps the following questions, publicized among those who use your international communication facilities, may challenge people to *think* before they lift the telephone receiver to call the international operator.

1. Does my message truly require discussion? Or am I conveying one-way information? Does the called person need to research my answer and call me back after I have given him my request for information? If the answer to either of the last two questions is "yes," a written international message should reduce cost.
2. Is it likely that the person I call will wish to talk about his problems as well as about my problem? Will such discussion prolong the call and increase my expense? Will the called person welcome an opportunity to have the cost of the discussion charged against the budget at the originating point?
3. How urgent is my discussion, request, or transmission of one-way information? Must the information be delivered personally to the recipient today? Would an international Telex transmission give even greater speed because dialing can be done quickly through the computerized facilities of the international carriers? Or is the expanding direct international dialing network giving similar or greater speed of connection to the points I wish to reach? If the message is not that urgent, could an inexpensive night cable or Telex message provide next-morning delivery?
4. Will the called person understand my language? If there is a lack of understanding, will this prolong the conversation? Would it be better to deliver my message in my language, or in the called person's language, in written form so that a written reply could be sent? Is there an advantage in having a written record of both ends of the conversation?
5. If the international telephone call is indeed necessary, is there any way that the call could be initiated between 5 P.M. and 5 A.M. in the United States originating area, so that the call would be billed at reduced international rates? Could the call be placed on a Sunday or to some countries on a Saturday to qualify for reduced rates?

If the international telephone call is necessary, here are facts which will help you make the call as easily as possible and at the lowest cost. You should also review the call-shortening techniques which were described in Chapter 1.

### Initiating the Call

Ask your telco operator to assist you. Perhaps you have found the distant telephone number on a letterhead or purchase order. If not, your operator will connect you to an international assistance operator, who will try to find the telephone number for you prior to placing your call. Almost all international calls are "operator-assisted." From certain exchanges in a number of cities in the United States, points in the United Kingdom may be dialed directly without operator assistance. Direct dialing is likewise possible to Puerto Rico, Alaska, the Bahamas, Hawaii, and the Virgin Islands. Direct dialing of calls is becoming more widespread. Your operator can inform you whether you can directly dial your call at a lower rate.

### Credit Card Usage

Your telephone credit card can be used for any international call. In certain foreign countries, your credit card can be used to call the United States.

### Collect Calls

Most countries will accept a request for a collect incoming call, and they will also accept your request to call the United States on a collect basis. In some countries, collect calls may be accepted only on a person-to-person basis.

### Time Periods

The local time in your city governs the rate schedule which will apply to the international call. In general, rates are reduced after 5 P.M. and on Sundays. In certain countries rates are also reduced on Saturdays. The United Kingdom DDD rate applies at all hours on a seven-day basis. In the Caribbean, Alaska, Hawaii, Puerto Rico, and the Virgin Islands, special rates are quoted for calls in the evening, night, and weekend time periods.

### Types of Calls

Calls can be placed on either a station-to-station or person-to-person basis, except in certain countries which will accept only person-to-person calls. The surcharge for person-to-person calls applies only to the first three minutes; overtime minutes carry the station-to-station rate. A substantial percentage saving results from using station-to-station calling, or, where available, DDD. The United Kingdom DDD rate saves 33 percent on a three-minute call over the operator-assisted station rate. There is a 14 percent DDD saving on similar calls to Puerto Rico and the Virgin Islands. The operator-assisted station call to other countries is billed at impressive savings over person calls. Examples of cost reductions are:

| | |
|---|---|
| Alaska | 42 percent |
| United Kingdom | 44 percent |
| Middle East | 25 percent |
| Australia, Japan | 25 percent |
| Mexico | 38 percent |
| Hawaii | 50 percent |

The experienced international communicator should be able to judge whether or not the odds favor the called person being present to receive the station-to-station or DDD call.

### Rates Applicable

I recommend that you ask your telephone utility marketing representative for a copy of the AT&T Long Lines department booklet, *International Telephone Rates*. This vest-pocket publication is issued periodically as rates change. It lists almost all countries to which international telephone service is available. Special rules are shown governing the use of credit cards or collect calling to or from a specific country. The booklet quotes daytime, night, and Sunday rates. Special rates are shown for direct dialing. Special rate periods, such as Saturday, are listed. If a country has different rates for different geographic portions of the country, typical rates are given. If the caller's location in the United States affects the rate of his call, a color coding arrangement lists an approximate rate from the general location.

Later in this chapter I will be quoting a few representative rates for full-rate cablegrams, deferred cablegrams, and international Telex messages; so that you can make comparisons more easily, I am presenting in Table 1 some representative international telephone rates as they existed during 1973. The federal excise tax applies to the rates shown.

When only a single rate is given in Table 1, the rate applies from any location in the 48 contiguous states to that country. To certain countries and United States possessions, the rate varies according to the United States location of the caller.

**TABLE 1 Three-Minute Daytime Rates for International Calls from the United States, 1973**

| Country or region | DDD rate | Operator-assisted rate: Station-to-station calls | Operator-assisted rate: Person-to-person calls |
|---|---|---|---|
| Alaska | $1.55–$4.75 | $2.70–$4.75 | $5.20–$7.25 |
| Australia | No | $9.00 | $12.00 |
| Bahamas | $2.40–$4.65 | $2.85–$5.40 | $4.50–$7.50 |
| Brazil | No | $9.00 | $12.00 |
| Central Europe | No | $6.75 | $12.00 |
| Hawaii | $3.00–$3.60 | $3.90–$4.65 | $7.80–$9.30 |
| India | No | No | $12.00 |
| Japan | No | $9.00 | $12.00 |
| Mexico | No | $2.25–$3.85 | $3.65–$6.60 |
| Middle East | No | $8.00–$9.00 | $12.00–$15.00 |
| Puerto Rico | $4.05–$5.10 | $4.80–$6.00 | $ 7.50–$10.00 |
| South Africa | No | $9.00 | $12.00 |
| United Kingdom | $3.60 | $5.40 | $ 9.60 |
| Venezuela | No | $5.00–$8.00 | $ 7.50–$12.00 |
| Virgin Islands | $4.05–$5.10 | $4.80–$6.00 | $ 7.50–$10.00 |

## INTERNATIONAL TELEX

Your Western Union Telex or TWX machine, or a similar unit which you have rented or purchased from an independent vendor, or a teleprinter which has been furnished to you by one of the international carriers will provide you with written wire communication to hundreds of thousands of offices throughout the world.

If you have equipped yourself with tape-operated teleprinter equipment, as I suggested in Chapter 14, you will be able to dispatch your written messages worldwide at a speed of 66.67 words per minute. If you wish, you can list your Telex or TWX number in one or more of the international Telex directories published abroad.

International Telex obviates problems which would arise in voice telephone conversations because of accents or unfamiliarity with the language being spoken. Telex provides a written record of the message transmitted or received. If the distant office is open at the time your transmission is received, a two-way conversation is possible between the two machines. The two executives involved can either operate the machine keys themselves or stand behind their operators and dictate answers as the questions appear on the machine's platen.

Use of this service is expanding rapidly. All the international common carriers offer international Telex service. If the circuits of one carrier are busy to a particular country, the sender can dial the circuits of another carrier. In general, the major carriers have computerized their service and offer special features which save dialing time and operator time and expedite the connection to the distant point.

## Important International Telex Characteristics and Features

**Coverage.** Worldwide.

**Speed.** 66.67 words per minute if tape transmission is used; otherwise the speed is determined by the skill of the operators involved.

**Equipment Availability.** In certain cities, the international carriers may be willing to furnish a teleprinter which will be connected directly to that particular company's dispatching facilities. In 1973 these cities were New York, Washington, and San Francisco. In other cities, teleprinters connected to Western Union's Telex, TWX, or Info-Com networks can be used for access to international circuits.

**Data Capability.** International Telex machines utilize the three-row teletype which produces the five-level Baudot code on the tape-cutting mechanism. When an incoming international Telex transmission has data implications, the receiving machine should be equipped with the mechanism that permits automatic tape punching as the transmission is received. If ASCII data language is required, the five-level Telex tape can be processed through a tape-to-tape converter to secure the ASCII tape for the data processing department.

**Rates.** Each international carrier will provide you with a rate schedule for international Telex. Also, a "break-even point" tabulation is available which tells you when an international Telex message will be less expensive than a full-rate cable or a deferred cable. In 1973, 49 destination countries could be reached at a cost of $2.55 per minute or less; 77 at $3 per minute; 51 at $4 per minute; and 3 at $5 per minute. Fifty-seven countries required a minimum conversation of only one minute; the balance required a minimum conversation of three minutes. The federal excise tax applies to international Telex transmissions, just as it does to domestic transmissions. Rates are uniform throughout the 48 contiguous states, regardless of the sender's geographic location.

**Hours of Availability.** International Telex, like its domestic counterpart, is an unattended service. Messages can be dispatched at any hour of the day or night, and they will be received automatically at the distant machine. When that office opens the following day, the messages will be ready for distribution.

**Directory and Information Service.** Most of your international correspondents will be anxious to notify you of their Telex number. All international carriers offer directory assistance; be sure you follow the procedure listed in their instructional brochures. Letterheads and purchase orders may list the distant Telex number. If you are using your TWX machine to seek Telex directory information, be sure that your operator understands the correct procedure, so that you will not be charged for the "information" call.

**Dialing Mechanics.** Dialing is done on the teleprinter by either a rotary dial or a key-pulsing mechanism resembling a ten-key adding machine keyboard. Automatic dialers can be used. Receiving equipment in the offices of the selected international carrier has been computerized. If the distant machine is not occupied, a connection can usually be established in less than 10 seconds. If all channels on the most direct route to the destination are busy, the computer may provide connection over an alternate route.

The cost of the Telex or TWX call from your office to that of the international carrier is absorbed by the carrier if you use the four-digit Telex number or the "800" TWX number which is listed in the Western Union directory for each international carrier. If no-charge numbers are busy, other Telex and TWX numbers are provided on a chargeable basis.

In 1973, 71 destination countries could be dialed directly by the originator of the call in the United States. Calls to other countries are completed with the assistance of the operators employed by the international carrier of your choice. Some carriers offer abbreviated dialing procedures as described in the next section.

**Special Services.** As international Telex is provided by all major international common carriers, there is much competition between them for your business. Each carrier has developed special services which may expedite your calls and save the time of your dispatching operators. Here are a few of the services described in the carriers' advertising literature. Each carrier will be glad to send you a complete package of their advertising and instructional brochures, and you should request such packages if you are sending cablegrams or international Telex messages.

1. *Abbreviated dialing.* By pressing one or two keys on the teleprinter, the number you wish to call is retrieved from the memory of the international carrier's computer, and the computer dials all necessary digits to make the connection. Thus, errors are avoided and operator's time is conserved on making connections to those you call frequently.

2. *Redialing busy numbers.* If the teleprinter reports "OCC," which is the international Telex symbol for a busy machine at the distant end, the subscriber can type a code symbol which instructs the computer to continue to dial the number until the message is delivered.

3. *Retrying busy circuits.* If the teleprinter reports "NC," which is the international Telex symbol for a busy circuit to the overseas point, the subscriber can instruct the computer to camp on the circuit. One international carrier's camp-on tests the busy circuit every 45 seconds for a 7-minute period before returning the message to the originator.

4. *Message retrieval.* It is possible to lose messages if a teleprinter should run out of paper during a period when it is unattended. The international carriers provide assistance in retrieving such messages. This service is also helpful if a message is misplaced in your office.

5. *Computer handling.* The carrier's computer will accept a message if the subscriber has encountered busy circuits or a busy number. The computer will deliver the message whenever circuits are available and the busy number is free.

6. *Itemized billing information.* Some of the international carriers provide a message number, the number of minutes of conversation, and the amount charged for each Telex call. This information is typed automatically by the computer at the end of each transmission. The monthly invoice rendered by the international carrier itemizes each call made and likewise shows the number of minutes and cost for each call.

## Suggestions for Using International Telex Efficiently

**Appoint and Train a Message Editor.** It is vital that messages to international points be understandable. The person reading the message may have only a limited comprehension of English. Clarity must not be sacrificed for brevity, but at the same time wordy, lengthy sentences may cause difficulties at the distant point. We recommend to multinational companies that their message center or international department designate someone who is familiar with the organization's technology and products to be the international communications specialist. This person screens every outgoing Telex and cablegram message and is given the authority to question unclear messages and ask the originator to clarify or rewrite. When a transmission is needlessly wordy, and when simple editing can be done to improve clarity and reduce transmission cost, that work is performed with or without the knowledge of the originator. We have found that a capable message editor saves several times his or her salary each month.

**Check All Cables for Telex Delivery.** As mentioned previously, the international carriers will provide information on publishers of international Telex directories. The carriers' assistance operators will likewise try to procure a specific Telex number for you. I suggest that you instruct your message editor or message center supervisor to be sure that a cable is checked against the international Telex directories before being dispatched at the higher rates usually applicable.

**Provide Proper Security for Your Teleprinter.** As international Telex is an unattended service, messages will be received automatically when your offices are closed. If messages contain confidential material, the teleprinter location should be locked and should not be accessible to those working after hours or to maintenance personnel.

**Supervise Use of TWX Equipment.** The international carrier's use of computers ensures that a message which is formatted incorrectly will be corrected before dispatch. However, the four-row TWX machine has certain keyboard differences from the standard domestic Telex teleprinter and from the machine furnished directly by the international carriers. These differences are explained in the domestic Western Union Telex-TWX directory, as well as in the instructional literature available from each international carrier. If you are using TWX for accessing the international circuits, be sure that you secure this information. Then make sure that your operators understand the keyboard conversions which they must make so that the correct message is received abroad.

**Check Cable Break-even Point.** If you are able to use a full minute of international Telex time, your cost per word for the 66.67 words transmitted will be about 4 cents in the $2.55 rate zone, 4½ cents in the $3 zone, and 6 cents in the $4 zone. In contrast, the rate per word for a full-rate cable in 1973 ranged from 21 to 34 cents.

The deferred "letter telegram" (LT) cablegram rate is 50 percent of the full rate, so the cost per word would range from 10½ to 17 cents, with a 22-word minimum charge.

From this listing of rates, it would seem obvious that international Telex would always be less expensive than any type of cable transmission. This is not necessarily true. If a message is to be sent to a point in Australia, for instance, it will be less expensive to send a full-rate cable if the message contains less than 32 words. A deferred cable will be less expensive than international Telex if it contains less than 64 words.

Thus, you can readily see that some supervision is required in your message center to ensure that words are counted before a decision is made on routing a message via international Telex or cable. A capable supervisor will likewise wish to consider the time urgency of the message. It may be helpful to your message center supervisor if you develop a specialized form to be used for both domestic and international written wire transmissions. The originator does not specify the routing method, as this is left to the expertise of the message center people. However, the originator *does* specify whether delivery must be made today or if delivery tomorrow or on the second day is satisfactory. Certain cablegram transmissions may be delivered on the second day when the International Date Line is involved.

**Review Message Header and Signature Requirements.** The "header" is the introductory material which appears first in the international Telex message. Each international carrier is glad to instruct you and your operators on formatting the message in a way which will result in prompt acceptance by the computer and speedy dispatch. Be alert for unnecessary words and figures in the header and in the signature. International Telex is a timed service. Keystroke elimination saves money.

**Review Domestic Telex Operating Suggestions.** I have given you suggestions on reducing the cost of domestic Telex and TWX transmissions in Chapter 14. If you will review those ideas, you will find that a number of them apply to your use of international Telex.

**Study Listing of Direct Circuits.** Each international carrier will furnish a listing of the countries to which they have international Telex circuits. Your message center supervisor should train your dispatching operators in second and third choices between carriers if the selected carrier reports a busy circuit condition. It may be better to test the circuits offered by several carriers rather than use any offered camp-on service via computer, particularly if your message is urgent.

**Avoid Machine-to-Machine Conversation via Keyboard.** Once an overseas connection is established, the receiving machine can talk to the sending machine. If people on both sides of the conversation wish to converse in writing, their operators will be typing manually as the people dictate the conversation. I estimate that the maximum speed for such manual conversation is about 30 words per minute. Thus, compared to the tape-operated Telex speed of 66.67 words per minute, manual typing becomes very expensive. The time of the transmission is more than doubled. If much conversation is involved, it may be better for the originating person to state the questions to be discussed, then disconnect the transmission. The receiving person does the needed thought and research. A new connection is then dialed from the receiving point to the originating point. A tape has been prepared with answers to the questions. The resulting comment or remaining conversation will then be shorter and less expensive.

If the receiving machine has traffic ready for the sending machine, the operator at the distant end can easily send her tape or tapes without disconnecting from the circuit. However, charges for the message being sent by the distant point will appear on the originator's bill.

**Scan Incoming International Telex Traffic.** The message center supervisor or message editor should observe the operation of teleprinters when those machines are receiving transmission from abroad. Is the printing slow and hesitant, with gaps of a few seconds? This indicates that the distant office is transmitting manually, rather than using tape transmission. It is possible that the office is not equipped for tape usage. This calls for prompt corrective action to replace the distant machine with a tape-operated machine. On the other hand, the operator at the distant point may not realize that tapes should be prepared before the connection is dialed. The editor likewise has an opportunity to observe wasteful messages and to take steps to educate the distant originators and operators in concise writing and key-stroke conservation techniques.

**Secure Telex Circuit Listings from FCC.** The FCC, Common Carrier Bureau, International and Satellite Communications Division, publishes a periodic report entitled *International Teletypewriter Exchange Service Between the Continental United States and Foreign or Overseas Points.* This publication shows which carriers operate international Telex service to the various overseas destinations. It may simplify your work in arranging a list of the carriers to use for reaching a specific foreign point.

## INTERNATIONAL TELEGRAMS

In the terminology of international telecommunications, the "cablegram" is known as an international telegram. For brevity, however, I shall use the word cablegram in this section, even though the cablegram may be transmitted by satellite or radio.

Two types of cablegram service are available in the United States. The full-rate cablegram is given priority by all international carriers, and delivery will be expedited at the distant point. The "letter telegram" is a deferred cablegram. The initials "LT" are placed just ahead of the first word in the address. Delivery is made as soon as possible during the following business day in the destination country, or on the second day in some countries which are separated from us by the International Date Line.

The full-rate cablegram has a minimum charge based on seven words. The LT cablegram has a minimum charge based on 22 words. The rate for the LT cablegram is always 50 percent of the rate for the full-rate cablegram. If a cablegram is not marked "LT" before the first word of the address, it will be sent full rate.

Cablegrams contain component parts consisting of the recipient's name, the street address, the city and country, the routing symbol showing your choice of international carrier, the text of the message, and an *optional* signature. As will be mentioned later, it is vitally important that you become familiar with the circuits offered by the various international carriers to the points where you are sending cablegrams. In general, dispatch your cablegram by the carrier which lists the greatest number of circuits to the distant point. Do not use a carrier which must relay your cablegram via a connecting carrier to reach your destination. This information is available from all carriers. Your outgoing cable must carry a routing indicator if you are to receive expedited service. The indicator ("Via RCA") (or ITT, WUI, TRT) must appear immediately to the right of the last address line.

Cable rates in 1973 varied from 21 to 34 cents per word for a full-rate message and from 10½ to 17 cents per word for a deferred message, subject to the minimums mentioned above. With some exceptions which will be mentioned later, every word in a cablegram is counted, including address, signature, and the "LT" routing symbol. The federal excise tax does *not* apply to international cablegrams.

A uniform rate applies per word to *any* destination within the foreign country or to a United States state or possession served by the international communications carriers. The rate includes the cost of carrying the message from the originating city to the dispatching point and for delivering the message from the receiving point to the destination address.

### Some Brief Suggestions on Sending Cables

The rules governing the sending and delivery of cables are set by an international commission. The word-count rules, particularly, are quite complicated. I will explain them in a later section of this chapter. Here are some suggestions which can be read quickly:

**Use LT Service.** If you are sending your cablegram to a business address, check the international time charts available from the international carriers. Determine the time in effect at the distant office, allowing one to two hours for delivery of your cablegram from the time you dispatch it to the international carrier. If you believe the distant office will have closed by then, put the letter telegram indicator "LT" in front of the first word of the address. Your cablegram will be delivered during the morning of the next business day in the distant country. In some cases, delivery is made in the afternoon of the following day or during the morning of the second day. The LT message reduces your cost by 50 percent.

**Use Cable Addresses.** These one-word addresses are registered in the distant country so that the receiving clerk can look up the code word. It will supply the addressee's full name, street address, city, and province or state. The use of the cable address code saves many chargeable words in your cablegram. You may find such an address on letterheads or purchase orders.

If your organization is contacted frequently by organizations located outside the United States, you should register your own cablegram address. Ask your Western Union domestic marketing representative for details, or request assistance from sales personnel in the offices of the international carriers. In 1973, Western Union quoted a yearly cablegram address rate of $15 for organizations located in New York City, Washington, D.C., and San Francisco. The rate was $25 yearly elsewhere.

**Route Your Cablegram.** The routing indicator should appear immediately after the last line of the cable address. It is set in parentheses: (Via . . .). The initials of the selected international carrier appear following the word "Via."

**Avoid Use of Punctuation.** If punctuation is required, the word-count rules specify that the punctuation mark must be spelled out: "period" or "stop," "comma," "question mark" or "quark."

It is best to compose your cablegram in natural thought groups which can be written in a continuous series without punctuation marks. Be *sure* your cablegram is understandable, however.

**Study the Run-Together Word Rules.** Proper names, compound street, building, and city names, and arabic numerals expressed as words may be less expensive when run together in a continuous letter series. See the examples quoted later in this chapter.

**Don't Use Spaces if a Dash Will Do.** If you are writing technical specifications, dimensions, and order or part numbers, use a dash to separate the portions of the number series, not spaces. The word count will then apply to the number of characters in the continuous series. Any space will indicate that the word or character group which follows will be counted separately.

**Eliminate Useless Words.** In Chapter 14, and again in the preceding section of this chapter, I have stressed the desirability of concise writing in messages which are subject to a cost per minute or cost per word. Any message originator can easily be taught principles of concise writing in an hour or two. Try eliminating connectives like "and." Use the "ing" form of the verb, like "shipping," instead of using two words "will ship." Unless a sensitive marketing situation calls for special courtesy, consider dropping words like "please."

**Simplify or Eliminate Signature.** Each delivered cablegram carries an initial "header" line which gives the city of origin, the date, and the time the cable was received by the international carrier. No charge is made for that information. Your recipient may immediately recognize your identity by the originating city and the content of your message. You are *not* required to sign the message. If you do sign it, use your own one-word cable address, or perhaps your surname standing alone, or your company name in as few words as possible. Try to eliminate titles, department names, first names, and initials unless they are needed to distinguish between two different persons.

**Consider Adopting Code Language.** If you are geographically located in a time zone which can communicate with your cable recipients on a same-day basis, you will be using some full-rate cables. Code language is permitted in full-rate cables, but not in LT cables. The international carriers can give you information about the availability of cablegram codes which you can purchase. Or you can make up your own code, being careful to limit the code words to five characters. Example: MRNCD might mean "We are shipping your order tomorrow." It would be counted as one word as compared to the six-word phrase.

**Become Familiar with Special Cable Services.** By using a chargeable code prefix before the first word of your cablegram message, you can instruct the delivering cablegram carrier to undertake special services which may be quite helpful to you and your recipients. A list of those services appears later in this chapter.

**Consider Use of Airmail.** Jet transportation of airmail is providing two-day service between the United States and many major international cities (see Chapter 13). If your recipient's office closes on Saturday and Sunday, perhaps an airmail letter or an air-letter sheet will convey your message at a fraction of cablegram cost. It should arrive on Monday—or you may affix a special-delivery stamp to ensure that delivery is made on receipt at the distant point. Be sure to mark such envelopes "Expres," which is the international word for special delivery. Your post office can furnish special stickers to be used for your international special-delivery mail.

**Secure Complete Literature from Carriers.** There are four major cablegram carriers in the United States, plus a number of smaller carriers which offer facilities to only a few selected countries. Every carrier is glad to send you a complete packet of literature on circuits, rates, word-count rules, dispatching techniques, and special services. Time and rate calculators are available. The Telex/cablegram break-even point chart is an essential tool if your message center is to pick the least expensive dispatching channel. Here are the names and addresses of the four major carriers as they existed in 1973:

| | Routing symbol |
|---|---|
| ITT World Communications, Inc.<br>67 Broad Street, New York, N.Y. 10004 | (Via ITT) |
| RCA Global Communications, Inc.<br>60 Broad Street, New York, N.Y. 10004 | (Via RCA) |
| TRT Telecommunications, Inc.<br>Prudential Plaza, Boston, Mass. 02199 | (Via TRT) |
| Western Union International, Inc.<br>26 Broadway, New York, N.Y. 10004 | (Via WUI) |

**Learn Word-Count Rules Carefully.** In the following pages, I will give you detailed instructions on word-count procedures, plus suggestions on reducing the chargeable word count. When your cablegram has been condensed as much as possible, remember that any cablegram of 11 words or less should be transmitted as a full-rate cablegram. A message of 12 words or more will cost 50 percent less if transmitted as an LT cablegram. If your message is less than 22 words and if it is to be transmitted as LT, you may wish to improve clarity by adding words until you reach the 22-word minimum.

**Consider Use of Full-Rate Cablegrams to Pacific Points.** The International Date Line complicates the determination of what is "today" or "tomorrow" when you are cabling the Orient or Pacific points. If it is urgent that you receive a reply soon, you may wish to send a full-rate cablegram. The time charts available from the international carriers will help you determine your choice of service. Here are two lists of destinations which have deferred delivery of the LT cable:

Delivery after 2 P.M., local time, on day following dispatch:

| | | |
|---|---|---|
| Burma | Japan | Philippines |
| Ceylon | Korea | Singapore |
| China | Macao | Tahiti |
| Guam | Malaya | Taiwan |
| Hong Kong | Okinawa | Thailand |
| India | Pakistan | Vietnam |

Delivery after 8 A.M., local time, on second day following dispatch:

| | | |
|---|---|---|
| Australia | Indochina | New Caledonia |
| Formosa | Indonesia | New Zealand |

**Dispatch Your Cable Correctly.** Users of Western Union domestic Telex service can dial a four-digit number which will connect the sender to any of the four major cablegram carriers. There is no charge for this transmission. TWX users can dial a special "800" number to connect to the cable carriers. This is likewise a no-charge call. The Western Union inward WATS numbers can be used for sending cablegrams. The answering operators record the message, then prepare a tape for dispatch to the international carrier of your choice. If the no-charge circuits are busy, each international carrier will accept messages telephoned to their headquarters offices. Special telephone numbers are furnished by each carrier. The sender pays the cost of the telephone call. The carriers can also be reached by chargeable Telex and TWX numbers. If you use the Western Union inward WATS system, be sure to indicate the name of the carrier you wish to use. Be careful to spell all proper names and repeat numbers. Be particularly careful to follow the word-count rules in telephoning your letter and figure groups, as well as your addresses and signatures.

**Accounting for International Cablegram Cost.** As you will find when you study the following pages, it is difficult to count the words in a cablegram if there are many figure groups. If your message center supervisor will study this chapter carefully, that person will be able to count and price an outgoing cablegram at the time it is dispatched. I suggest that you check the priced copy against the billing which you will receive from the international carriers or from Western Union. Be sure that the number of chargeable words shown on your bill agrees with the number of chargeable words shown on your file copy of the transmitted cablegram.

**Secure Routing Information from FCC.** The FCC, Common Carrier Bureau, International and Satellite Communications Division, publishes a periodic report entitled *Normal Routing of International Message Telegraph Traffic Between the Continental United States and Overseas and Foreign Points.* This report shows the following helpful information:

Column 1: Country of destination
Column 2: Name of cablegram carrier
Column 3: Originating city in the United States
Connecting city abroad
Name of delivering carrier of government agency
Special information on hours of operation

Users of cablegram service may find this report helpful in planning the choice of international carriers for cablegram transmission.

## SPECIAL CABLE SERVICES

Any international carrier will give you full information on the many special cable services which will expedite transaction of your overseas business. The principal special service indicator symbols are shown and explained below. A charge is made for only *one word* for the symbol and its accompanying explanatory phrase, except where noted.

RP — (followed by a phrase like "DLRS5.50") Reply has been paid for by recipient up to $5.50. Any excess of cost of reply is collected from the foreign addressee. Any surplus remaining from prepayment is refunded to the sender in the United States when application is made either by the sender or the addressee within four

| | |
|---|---|
| | months from the date of voucher issue, but only if the difference is at least 65 cents. |
| NUIT | In most overseas countries, delivery of a full-rate message is not attempted after the early evening hours. This symbol requests delivery during the night, if possible. |
| JOUR | Delivery is to be made *only* during the day. |
| CHEZ | Use this symbol instead of the words "in care of" in the address. |
| FS | If you think your addressee may have moved or traveled to another address, this symbol ensures the forwarding of your message if the new address is known. Forwarding charges not collected from your addressee on each forwarding will be collected from the sender. You may furnish a series of addresses. |
| PC | This is used to indicate that you wish to know the date and time of delivery of your message. You will be charged seven additional words at the full rate for this notification. This service is not available for LT messages. |
| TF | (followed by your addressee's telephone number) Telephone the cablegram to the number indicated. |
| TLX | (followed by your addressee's Telex number) Deliver the message by sending it over your recipient's Telex receiver. |
| REMETTRE | (followed by a date) Deliver on a specific date. |
| MP or MAINSPROPRES | Deliver to addressee only. |
| GP or POSTERESTANTE | Hold in "General Delivery" at delivering government post office. |
| GPR | Hold for delivery as registered mail at delivering government post office. |
| PR | Forward to a certain point and deliver by registered mail beyond. |

| | |
|---|---|
| TR | Hold for pickup at destination telegraph office. |
| POSTE | Send to certain point and then deliver by regular mail. |
| PAV | Send to certain point and then deliver by air mail. |

## CABLE WORD-COUNT RULES

The international word-count rules are followed by all countries. If you will familiarize yourself with them, you can determine the cost of a particular cablegram before it is dispatched. Such a calculation is important when you must decide between dispatching a message as a cablegram or placing it on an international Telex circuit.

As the cost of a cable is based on the number of chargeable words, employees of Western Union and the international carriers must be carefully trained in the application of the international word-count rules. It is possible that a new employee might place an incorrect price on your cablegram. I suggest that you ask your message center supervisor or cablegram editor to count the words in selected cablegrams. Then determine the cost of the cablegram from the rate tables which are available from any carrier. Then compare your cost estimate with the actual billing you will receive. If there is a discrepancy in your favor, report the facts to the carrier so that your account may be credited.

### Addressing and Signing Your Cable

**Free Symbols.** Two portions of the cablegram address are transmitted free by all carriers, and need not be counted in your determination of word count:

1. The country of destination
2. The routing symbol

Thus, Paris(France) would carry a word count of one, not two. The phrases (VIA RCA) (VIA ITT) (VIA WUI) (VIA TRT) or similar routing designations are carried without charge.

**Service Indicators.** No symbol is necessary for full-rate service. If deferred service is desired, the initials LT should appear just before the name of the addressee:

LT FRANKLIN RHODES

If special service is desired, the service indicator should also precede the addressee's name:

PC ROBERT SEIDEL

A charge of one word is made for each service indicator.

**Cable Addresses.** You or your foreign addressee may register a special cable address which consists of one word. It is used alone. The second line of the address contains only the city and country of destination. If the message is to move by LT service, the indicator LT precedes the cable address:

LT LINCOLN
PARIS(FRANCE)

To ensure that your cable is counted correctly, be sure to have your operator omit the normal space between the city and country:

Wrong: PARIS (FRANCE)
Right: PARIS(FRANCE)

Any international carrier will furnish full information on the mechanics of selecting and registering a money-saving one-word cablegram address. If you wish, the address may be "reversible," meaning that it can also apply to your company offices in cities overseas.

**Street Addresses.** Always *eliminate* a street address if you are using a one-word cablegram address. The street address is not needed. If a street address is necessary, put it on the line next below the addressee's name. The city, county, and country should then be run together in one continuous phrase, with the country set apart by parentheses:

| | | |
|---|---|---|
| City: | Welwyn Garden City | Expressed as |
| County: | Herts | WELWYNGARDENCITYHERTS(ENGLAND) |
| Country: | England | |

If you have a compound street name in the address, run the phrase together:

RUEDELAPAIX (for Rue de la Paix)

London addresses should include the postal district:

LONDONSWI(ENGLAND)

**Signatures.** It is NOT necessary for *any* signature to appear. If your cablegram goes to your own office or factory, the sense of the message or the name of your addressee may be sufficient to identify the sender. Your city of origination will appear on the addressee's copy. There is no charge for this identification of the originating city.

If a signature is necessary, try to use your own one-word cablegram address or the surname only of the sender. Every word, including initials, titles, addresses, and department names, is chargeable when it appears in a cablegram signature.

International rules permit the same one-word cable address for a company in the United States and its subsidiary or office overseas. Such a word is called a "reversible" address.

**Punctuation.** The keyboards of cablegram transmitting equipment permit use of the following punctuation:

Period or decimal point (.)
Comma (,)
Colon or division sign (:)
Apostrophe or sign for minutes or feet (')
Hyphen, dash, or subtraction sign (-)
Parentheses ()
Fraction bar, stroke, or division bar (/)
Quotation mark or sign for seconds or inches (")

The following symbols must be spelled out in letters or a word:

$ - DOLLARS or DLRS
# - NUMBERS or NOS
& - AND
% - PERCENT or 0/0
¢ - CENTS

Here are the word-count rules for punctuation:

1. Punctuation used in its normal sense is transmitted *only when specifically requested by the sender.* It is charged for at one word for each punctuation symbol. Exceptions:

*a.* The two signs forming brackets (parentheses) and the signs forming quotation marks (inverted commas) count as one word when they enclose one or more words or groups.

*b.* A dash or hyphen used to separate a whole number from an associated fraction or to separate figures from an associated percent sign (0/0) is neither counted nor charged for.

*c.* A fraction bar in the address, used to form a house number composed of figures or of figures and letters, is neither counted nor charged for.

EXAMPLE: (APARTMENT 1252/C = 2 words
(12-0/0 = 1 word

2. Punctuation marks appearing in figure or letter groups count as one *character* each.

EXAMPLE: 125.75 - 2 words (6 characters)
12.57 - 1 word (5 characters)
12-9-73 - 2 words (7 characters)

## Words

**Length of Words.** Words of any language which can be found in a dictionary are counted as one word for each 15 characters.

INDEMNIFICATION - 1 word (15 characters)
INCOMPREHENSIBLE - 2 words (16 characters)

If your word has 16 characters or more, use a synonym.

**Compound Words.** Compound words found in a dictionary count as one word.

AIRMAIL - GOODWILL - COOPERATE

Compound words composed of two or more dictionary words will be counted in segments, even if run together. See later paragraph concerning using words for numbers.

| | |
|---|---|
| BLADING (for bill of lading) | = 2 words |
| AIREXPRESS | = 2 words |
| NINETYCENTS | = 2 words |

**Run-Together Words.** Short dictionary words which total five characters or less when run together are counted as one word in *full-rate messages only.*

CANDO - DONOT - TOBE

**Code Words.** Code words or secret language may be used only in full-rate messages. The count is based on one word for each five characters. If words are used in LT messages which cannot be found in a dictionary, they usually are counted in segments if a portion of the word is found in a dictionary. Be sure to register your company branch office with the foreign country's local Communications Bureau, or that government may reject coded messages.

| | In full-rate message | In LT message |
|---|---|---|
| MRNCD (Code) | 1 word | Not admissible |
| RETEL | 1 word | 2 words |
| RECAB | 1 word | 2 words |
| RECABLE | 2 words | 2 words |
| AMEX (for American Express) | 1 word | 2 words |
| AIREX (for air express) | 1 word | 2 words |

**Proper Names.** Proper names, organization names, and abbreviations in common use are counted as one word for each 15 characters, providing no space appears.

| | | | |
|---|---|---|---|
| UNESCO | = 1 word | DE MARCO | = 2 words |
| ALCOA | = 1 word | VANDEBRANDE | = 1 word |
| DEMARCO | = 1 word | VAN DE BRANDE | = 3 words |

**Compound Street Names.** Compound street names appearing in the address should be counted as one word whether or not spaces appear, but spaces must be eliminated in the text if such addresses are to be counted under the 15-character rule.

| | In address | In text |
|---|---|---|
| AVENUE OF THE AMERICAS | 2 words | 4 words |
| AVENUEOFTHEAMERICAS | 2 words | 2 words (19 characters) |

## Figures and Letters

**Figure Groups.** Subject to the exceptions shown below, figure groups are counted as one word for each five characters. All punctuation used in such groups is counted as one character. Dates should be expressed in figures if the recipient will understand.

| | |
|---|---|
| OCTOBER TWELFTH | = 2 words |
| OCTOBER 12 1973 | = 3 words |
| OCTOBER 12, 1973 | = 4 words (comma counts as one word) |
| 10/12 | = 1 word |

*Be sure* that your overseas correspondent understands that in American usage 10/12 means October 12, not December 10.

**Letter Groups.** Subject to the exceptions shown below, letter groups are counted as one word for each five characters. Many companies have developed their own simple codes consisting of five-character words:

MRNCD = We are shipping your order tomorrow.

Code or secret language is admissible only in *full-rate* messages. Companies interested in purchasing standard codes understood throughout the world should contact the international carriers to secure names and addresses of organizations which sell standard codes.

**Figures Written as Letters.** Figures written in letters are counted as one word for each 15 characters, providing no space appears.

| | |
|---|---|
| TWO HUNDRED AND THIRTY FOUR | = 5 words |
| TWOHUNDREDANDTHIRTYFOUR | = 2 words (23 characters) |
| TWOTHIRTYFOUR | = 1 word (13 characters) |
| 234 | = 1 word |
| TWOPERCENT | = 1 word |

**Mixed Letters and Figures.** Subject to the exceptions shown below, a mixed group is counted as one word for each five-character segment of its component parts. Letters and figures *may not* be mixed to achieve the five-character count.

| | |
|---|---|
| AB456 | = 2 words |
| ABCD123456E | = 4 words (4-character letter group, 6-character figure group, 1-character letter group) |

## Special Combinations

**Time.** Time expressed in letters and figures is counted as one word for each five characters providing no space appears.

| | |
|---|---|
| 10PM | = 1 word |
| 23H30 | = 1 word (Continental time) |
| 11:30PM | = 2 words (7 characters) |
| 23:30 | = 1 word |

If your recipient understands the 24-hour clock, it should be substituted for our 12-hour system.

**House Numbers.** House numbers, room numbers, and travel space appearing in the text or address which include both letters and figures are counted under the five-character rule.

| | |
|---|---|
| 15/B | = 1 word |
| 1021/A | = 1 word in address; 2 in text |

**Other Numbers.** Ordinal numbers, multiplication, and division are subject to the five-character rule.

2ND = 1 word
17ME = 1 word
1001ST = 2 words (6 characters)
15x6 = 1 word (indicating multiplication)
15 × 6 = 3 words (indicating dimensions)
15×6 = 1 word (spaces were eliminated)
15:6 = 1 word (indicating division)

**Commercial Terms.** Commercial marks, trademarks, abbreviations in general use, part numbers, technical terms, references, and indications are counted under the five-character rule if they are found in public catalogs, price lists, invoices, bills of lading, or similar documents and if no spaces appear within the series. Substitution of dashes for spaces may convert such disconnected series to continuous series at a saving in word count.

ABCD-123456-E = 3 words (13 characters)

**Money.** Sums of money are subject to the five-character rule if no space appears:

DLS50 (for $50.00) = 1 word

## TELEPRINTER LEASED CHANNELS

Organizations which use large quantities of international telecommunications should investigate the possibility of installing one or more teleprinter leased channels. In general, such a channel may become economically feasible at 1973 rates if you are sending 100 or more international cables or if you are using over 200 minutes of international Telex time to one particular point or area.

If you are communicating with your *own* offices overseas, it is important to calculate the cost and number of cables, plus the cost and minutes of Telex transmission which are being sent to you from those international points. You can then procure up-to-date rates for teleprinter leased channels from the marketing representatives of the international carriers.

The teleprinter channel is connected directly to your office or agency abroad. The channel is yours to use 24 hours a day and 7 days a week. No dialing is necessary. As soon as you press the button to start tape transmission, or commence typing manually on the line, the overseas machine will begin to print your message on its platen. If desired for data processing, tapes can be prepared on incoming transmissions at both ends.

Teleprinter channels are arranged for full-duplex operation. This means that half of the channel operates in the outbound direction; the other half operates inbound. As a result, a constant stream of information can flow in both directions at the same time. As the monthly cost is fixed, your cost per word declines as traffic increases.

Such a leased channel is private to your organization, eliminating the need for coding confidential messages. You establish your own priority systems for urgent transmissions.

The standard speed of the normal leased channel is 66 words per minute. However, many firms find that a "dedicated" circuit which works all day and all night can operate at a slower speed and still handle the needed volume. Consequently, all the international carriers will furnish lower-cost circuits operating at half speed (33 words per minute) and quarter speed (16 words per minute).

In considering the installation of a leased channel, ask yourself if you could concentrate traffic on a single circuit to one of your major offices abroad. Perhaps you might pick a circuit from New York City to Geneva. Then you could arrange for your messages to be relayed by European Telex to offices in other European and Middle Eastern cities. European Telex is very inexpensive; your messages can be relayed at a fraction of the cost you would pay if you sent a direct cablegram or international Telex message from your headquarters to the distant point.

You may also wish to consider the possibility of providing a direct connection to your overseas leased channel from your offices on your domestic wire network. If your other United States offices are sending cablegrams and international Telex messages abroad, tabulate the volume and cost of that traffic and determine whether a centralized dispatching point would produce savings.

All the international carriers provide assistance in determining if a teleprinter leased channel will save you money. The carrier will be glad to analyze your message traffic in both directions and will give you a cost study for the proposed system.

The international leased circuit requires several items of equipment. You can furnish the equipment or it can be secured from the international carrier. The typing reperforator has a typewriter keyboard on which the operator perforates a tape which will carry the message. The automatic transmitter sends signals out on the channel as directed by the perforations on the tape. The teleprinter monitor is connected to the automatic transmitter and prepares exact copies of the message in page form. The basic machine for receiving messages is a receiving teleprinter which prints incoming

**TABLE 2 Rates per Month of Some Teleprinter Leased Channels, 1973**

| | Standard speed (66 words per minute) | Half speed (33 words per minute) | Quarter speed (16 words per minute) |
|---|---|---|---|
| New York to Europe | $1850 | $1233 | $ 740 |
| New York to Latin America | 2400 | 1600 | 960 |
| San Francisco to Australia | 3160 | 2107 | 1264 |
| San Francisco to Japan | 3160 | 2107 | 1264 |

messages in page form. Additional units of this type can be installed in executive offices to duplicate incoming transmissions.

Experienced international communicators believe in "blind" tape creation. The operator types from copy without seeing the result of the typing on the machine platen. Such tape preparation has proved to be faster and more accurate, as the attention of the operator is not detracted by watching the printing on the page. However, a page-printing teletypewriter is available as optional equipment for those who prefer to read the message as the tape is being prepared. In busy communications centers, the various units of equipment may be combined into a teletypewriter set. A multiple transmitter is available which accommodates two message tapes and one numbering tape so that the machine automatically numbers each message for transmission. As mentioned earlier, a receiving reperforator can be secured which will perforate tapes of incoming messages. Such tapes can then be relayed to other points on the domestic wire network.

The cost of the telegraph-grade teleprinter leased channel has decreased steadily. Typical monthly rates as of 1973 are given in Table 2.

These rates include equipment required at the originating point. The carriers furnish information on equipment charges at the distant end. In general, the charge at the overseas destination is similar to the charges listed in Table 2. For instance, the rates per month for both continental United States and overseas terminals for certain circuits are shown in Table 3.

**TABLE 3 Rates per Month for Circuits Between Continental United States and Overseas Terminals, 1973**

| | Standard speed | Half speed | Quarter speed |
|---|---|---|---|
| San Francisco and Hawaii | $2640 | $1760 | $1056 |
| New York City and Hawaii | 3850 | 2556 | 1526 |
| New York and Puerto Rico | 1740 | 1160 | 696 |

## VOICE/RECORD LEASED CHANNEL SERVICE

In the preceding paragraphs I have discussed telegraph-grade circuits. These circuits have a relatively narrow bandwidth. If the bandwidth is increased, it is possible to convey information or data at higher speeds. A voice conversation can be held. Facsimile transmissions can be sent. Normal telegraph-type traffic can also use the circuit.

In the typical voice/record (V/R) circuit, equipment is provided which will convert the circuit to a different use by the touch of a button. If needed, the V/R circuit can be used for four simultaneous telegraph-type transmissions at the standard 66 word per minute speed or at 100 words per minute.

Some organizations may not require telegraph-type transmissions, but need voice and data communication with points abroad. This need is met by alternate V/R service.

Another service is known as interpolated V/R service. As a person pauses to catch his breath while talking internationally, the equipment silently transmits bursts of high-speed data. When the line is not being used for voice conversation, data are transmitted at an even higher rate. Facsimile transmission is possible, and this service likewise can be divided into four standard-speed teleprinter channels. The service provides two-way data service while the voice conversation is going on. Standard teleprinter service may likewise continue while the circuit is being used for voice or data.

If your organization has sophisticated data processing equipment which requires a transmission speed higher than the 2,400 bits per second possible with standard V/R circuits, the international carriers can furnish V/R circuits of higher capacity. Data transmission could be sent in 1973 at speeds of up to 60,000 bits per second. Such circuits can also be divided into as many as 22 standard telegraph-type channels at a standard speed of 66 words per minute. If higher-speed printers are required, from 100 to 1,000 words per minute, the subdivision of channels can be adjusted to accommodate the requirements of these machines.

In 1973, the rate for the United States terminal of a V/R circuit from New York to Europe was $4625 a month. The international carriers provide information on terminal charges at the distant end. A V/R circuit between New York and Hawaii, including terminals at both ends, cost $8880 a month in 1973. A similar circuit between San Francisco and Hawaii cost $6600 monthly.

## EXECUTIVE HOT LINE V/R SERVICE

In Chapter 14 I described the Western Union Hot Line service available between certain cities in the United States. Some of the international carriers are offering an almost exactly similar service. The charge is based on a $\frac{1}{10}$-minute increment. Instruments are provided for executives at their desks. No switchboard connections are necessary. When the telephone receiver is lifted, the instrument on the distant desk begins to ring. Exclusion features are available to ensure privacy. Data transmission is possible at speeds up to 2,400 bits per second, with appropriate data terminals and modems.

One carrier offered this service in 1973 between New York and San Juan and between New York and Brazil. The Puerto Rican rate was 18 cents per six-second increment; the Brazilian rate was 40 cents per six-second increment. However, the carrier required a minimum of $2400 yearly in traffic from each end of the Puerto Rican circuit, and the Brazilian circuit required a $7200 yearly minimum from the United States, plus $7800 yearly minimum from Brazil, plus a 40 percent Brazilian government tax. By the time you read this paragraph, the Hot Line network will have been extended by the carriers. Their marketing representatives can give you current information.

## DATEL SERVICE

If you have data to exchange with a point overseas, and the amount of data does not justify your acquiring a full-time leased circuit, Datel service may be very helpful to you. It is offered by all the major international carriers. It is charged for by the minute. Speeds range from 100 to 1,200 bits per second.

If you are located in a "gateway" city (New York, San Francisco, or Washington), the international carrier may provide you with a voice-grade private telephone line between your office and theirs. Your data flow through this line into an international V/R circuit. A monitoring telephone is furnished for voice control of the Datel call. The international carrier contacts the overseas operator, who dials the telephone number of the distant computer or office. When the called number answers, you are instructed to proceed with your call. The distant terminal may be arranged for unattended operation.

If you are located in another city, you must acquire a private voice-grade line to the gateway city closest to you. If you are a Western Union broadband subscriber (see Chapter 14) you can call the international carrier on your broadband circuit and be connected to the overseas channel. In 1973, there was a three-minute minimum which applied to Datel service. Costs per minute ranged from $1.25 to $2.10 for Hawaiian calls to $3 for Puerto Rico and $4 for European points.

### PHOTO-TRANSMISSION SERVICE

The word Wirephoto frequently appears under a newspaper picture. Photographs and drawings can be transmitted abroad by the international carriers. The service is available only in certain of the gateway cities. Material from other cities must be sent to the carrier for transmission. Typing must be in 8-point type and preferably in 10-point. In general, 8½ by 11-inch copy is preferred; larger copy is transmitted in sections.

There is a minimum charge for this service based on a transmission of 23.3 square inches, with an additional amount charged for each 15.5 square inches. In 1973 the minimum charge varied from $20 to over $40, depending on destination.

### CLOSED CIRCUIT TELEVISION

Certain carriers offer to prearrange a two-way closed circuit television conference, either in a special studio or in the offices of the executives involved. Rates are quoted on request.

### COMPUTER SWITCHING FOR INTERNATIONAL AND DOMESTIC MESSAGES

When an organization has a domestic wire network, and also receives messages from abroad which must be switched to the domestic offices, the operators in the message center may physically tear the tape from one receiving teleprinter and carry it to an outgoing teletypewriter for dispatch. In larger organizations, the message switching may be done by an electromechanical arrangement. This machine reads coded destinations and places the message on the appropriate circuits. Certain of the international carriers offer a minicomputer switching system on a shared-time basis. The leased channels formerly terminating in the message center are connected to the international carrier's computer. Automatic numbering is provided, in sequence, as messages are sent to each station on the network. There are different levels of message priority. Automatic alternate routing is provided for circuits that are out of order. Daily traffic reports can be secured showing the number of messages received from and sent to each point on the network. Further details can be provided by the carriers.

## MARINE SERVICE

The international carriers operate marine departments which maintain complete data on the registry, location, and destination of all ships at sea. The service is available on a 24-hour, 7-day basis. Shore to ship service utilizes full-rate international telegrams (cablegrams). Charges in 1973 were $0.3325 per word to ships of American registry and $0.3825 per word to ships of foreign registry providing that transmission was accomplished from shore to ship by one of the United States coast transmitting stations. A seven-word minimum charge applies, and word-count rules follow the international pattern described earlier in this chapter. The name of the coast station counts as one word. If the sender wishes notification of delivery, the indicator "PC" is used preceding the first word of the address. The additional charge in 1973 for PC service was $0.5425.

From ships at sea, three services are offered. All messages are transmitted to the coast station by radio. A full-rate message is transmitted by telegram from the coast station to the addressee. The night letter will be delivered to the addressee the following day. The sea letter will be forwarded to your addressee by first-class mail or airmail from the coast station.

CHAPTER SIXTEEN

# ACCOUNTING FOR TELECOMMUNICATIONS EXPENSE

Management asks: How many clerical expense dollars should we spend on allocating the cost of using our telecommunications services? If we make detailed charges to departmental budgets, will our managers inspect those charges and take steps to investigate variances and reduce costs?

A cost-conscious company does not wish to increase clerical overhead expense unless the contemplated departmental or individual cost allocation will indeed produce an increased awareness of cost and a corresponding cost-reduction effort. Here are some suggestions that will help you appraise the accounting effort which you may wish to apply to the telecommunications portion of your administrative expense control program.

## PERMISSIVE VERSUS CONTROLLED ENVIRONMENTS

In many organizations a department head is given autonomy to operate his department in a way which will produce the performance required by his management's objectives. He is free to contract for the peripheral services needed to implement his effort. He may decide that a multibutton telephone, a hands-free attachment, and a departmental WATS line are all required to bring his department's profit contribution up to management's requirements. Under these conditions, the *permissive* environment fully justifies the charging of incurred costs to the using department or division.

In other organizations the department head may have full authority to control costs for labor, material, and needed departmental overhead, but management reserves the right to control the furnishing of central administrative services. If additional telecommunications service is required, the department head makes that request of the administrative services manager. The request is checked against any standards which have been established. If it meets the standard, it is approved. If it does not, or if the service requested is not covered by a standard, the administrator visits the department to provide expert advice and counsel on solving the problem. Under this *controlled* environment, company management may feel justified in concentrating certain telecommunications costs in one account. This is because proper controls have been established, and new services cannot be ordered without expert advice and management approval. Clerical and data processing costs are reduced by eliminating the detailed cost allocations.

## PRORATED EXPENSE ALLOCATIONS

To avoid the clerical and data processing costs involved in detailed cost accounting, some managements allocate telecommunications cost on a prorated basis. Equipment cost may be charged to a department according to the total number of people using telephone instruments, or perhaps according to the number of instruments allocated to that department. A variation of this procedure is based on an actual cost analysis, in detail, performed once or twice a year. Rather than allocating each cost item monthly, management establishes a percentage distribution which is based on the detailed cost analysis. That percentage distribution is used until the next detailed audit is performed.

Many organizations of considerable size have decided that the cost of every administrative service must be charged to the department which uses that service. While a prorated expense distribution can be used for this purpose, it is difficult for departmental managers to agree that they should pay for a sudden or seasonal increase in service cost which is incurred in another department, without a corresponding change in the prorated percentage basis. Consequently, unit costing is used whenever the cost of a specific service can be identified and charged to the user.

## COST-ACCOUNTING PROCEDURES FOR TELECOMMUNICATIONS SERVICES

Let us consider how the various methods of accounting may be used for each of the common telecommunications services.

### Long Distance Toll Charges

As the monthly toll statement lists the cost of each telephone call, it is possible to charge that cost against the user if the user's identity is known. As mentioned in Chapter 1, this identification is established by toll ticketing. The toll ticket may be prepared by the user when the call is dialed, by the operator at the switchboard, or by automatic toll-ticketing devices associated with the telephone switching equipment. Centrex systems provide such a cost allocation automatically for each numbered Centrex line. If manual ticketing is used, the tickets usually are sorted by date. They are compared with the telephone bill so that charges can be inserted. Appropriate taxes are added, and the accounting department makes the charge against the departmental budget, perhaps using a punched card procedure. When a toll call cannot be identified, telco's business office is asked for the identification of the called telephone. If a person-to-person or operator-handled call is involved, telco's records will provide additional information which may help allocate the cost. In general, if users may make toll calls at their discretion, it is wise to have their budget reflect the cost of those calls.

### WATS and Other Dedicated Lines

WATS calls cannot be identified specifically from telco records, as the WATS tariff does not require telco to list individual WATS calls. While telco may be willing to provide detailed printouts of WATS calls once or twice a year (see Chapter 3), the administrator wishing detailed WATS unit cost information must originate his own method of securing that cost. If WATS lines are procured by dial access, there is no way that the length of the call can be determined, so WATS charges may be distributed on the basis of the number of telephones in the department which have access to WATS. Or, if some toll calls are made to areas not covered by WATS bands, the WATS cost may be distributed proportionately to toll cost. If WATS is under operator control, and if that control is designed to limit lengthy calls, the administrator may place all WATS costs in one account. In some companies, those who are willing to wait for WATS are relieved of budget responsibility for

WATS calls. Toll calls bypassing WATS circuits are charged to those who were unwilling to wait. This type of an incentive may indeed help the administrator secure better cooperation from callers. WATS usage should increase; toll calls should decrease. The same kind of accounting procedure may be used for calls made over tie lines and FX circuits.

As you learned in Chapters 3 and 12, interconnected equipment may be used to bring automatic toll ticketing to your organization. The printout associated with the equipment will indicate the station line number or the identification number of the individuals who originated the call.

## Message Units

If an organization is unwilling to restrict certain telephones so that the user cannot secure an outside line for dialing local calls, it may be advisable to prorate message unit charges on the basis of telephone stations which are unrestricted. If restriction has been carefully worked out, and if management feels that the remaining lines are being used chiefly for company business, it will be more economical to adopt lump-sum accounting for message unit expense.

## Western Union Domestic Telex

Unless Telex messages are lengthy, the normal cost per message may not justify an expenditure for cost allocation on a per-message basis. If the message center is well supervised, the dispatching operator will be eliminating key strokes to reduce transmission cost. The message center supervisor will be doing basic editing to eliminate unnecessary words before the transmission tape is cut. When the message center is staffed by trained personnel, it may be best to account for Telex transmission and equipment costs on a lump-sum basis. The costs listed for typical transmissions in Chapter 14 are not great enough to justify detailed cost accounting on a per-message basis.

## Western Union Domestic TWX

As you learned in Chapter 14, TWX costs are usually higher than Telex costs, and Western Union's monthly TWX toll statement itemizes each TWX call. Thus, it is possible to allocate message costs on the per-message basis if desired. However, as those costs range from only 20 to 70 cents per minute, and as typical transmissions are short, the unit accounting cost may not be justified.

### Western Union Telegrams (PMS)

When a day telegram now costs a nimimum of $3.95 without physical delivery and an overnight telegram costs $3.20, it is evident that other methods of message transmission are less expensive. The administrator may choose to allocate telegram costs directly to the sender in order to reduce the use of telegrams.

### Western Union/Postal Service Mailgrams

As Mailgram service is proving to be very popular, large quantities of these messages may be used by departments handling sales and collections. Because the Western Union computerized billing lists every Mailgram, showing destination and cost, unit accounting can be done if the average cost per message seems to justify the accounting cost involved.

### Private Line Networks

Voice circuits are very expensive; telegraph circuits are less expensive, but either network will produce a substantial annual bill. If the administrator periodically evaluates the traffic passing over the network, he can use the resulting data to prorate network cost among the users. Character counts are excellent sources of cost accounting data. The telcos can provide traffic metering equipment. Western Union Info-Com users (see Chapter 14) secure traffic information automatically.

## SOURCES OF ACCOUNTING INFORMATION

If costs are to be charged back to users, the administrator must have dependable sources of input information on those costs. Here are a few suggestions on sources.

### Monthly Telephone Bill

Despite the many differences in accounting procedures among the utilities, the monthly telephone bill almost always lists these cost classifications:

1. *Local service.* This is the lump-sum billing for the rental of your telephone equipment.
2. *Message units.* Most bills give the number of message units, as well as the chargeable amount. By dividing the number into the cost, you can determine the current rate per message unit.
3. *Calls and telegrams.* This category totals the amount billed for long distance toll calls. If any of your people telephone telegrams to Western Union's local telephone number and charge those

telegrams to your main number, the cost of the telegrams will be included among your toll calls.

4. *Other charges and credits.* This classification includes all charges for installing, moving, or changing telephones. It also includes credits for the unused prepaid rental of telephones disconnected and charges in advance for telephones added to the system between standard billing dates.

5. *Directory advertising.* You are entitled to one white page listing and one yellow page listing for each PABX system rented and for each supplementary business line which you may be using to supplement your PABX system. These listings must be in plain type, not boldface type. If you require supplementary white page listings in ordinary type, their cost will be included under "local service," and you will find the total cost for all listings in the inventory of equipment which telco will send at your request. If you require boldface listings in either the white or yellow pages, the cost of those special listings will be under "directory advertising." Your display advertising in the yellow pages is likewise billed under this title. If you wish to secure details of the charges which are lumped together under "directory advertising" you must call or write the special telco department which administers this function. Directory advertising is always covered by a contractual document. You may find that your advertising department has a copy of the contract which you can inspect.

6. *Taxes.* Some telcos show federal, state, and local taxes as an item to be added to the total of your bill; other telcos show taxes as part of each item of billing. If your taxes are lumped at the bottom of your bill, you may have to secure a tax breakdown from telco so you will know which taxes relate to which items on your bill.

## Other Sources

7. *Toll statements.* Every telephone bill lists toll calls by date. Most telcos are now identifying the *type* of call by a number or letter code. Thus, by inspecting your toll bill you can determine which calls are person to person. Also, separate codes are used for third-number, collect, and credit card calls. Directly dialed calls are identified. Centrex statements are issued for each numbered Centrex line; thus toll costs are itemized for that line, simplifying cost distribution.

8. *Message unit billings.* In certain areas, such as California and portions of Illinois, telcos are itemizing message unit calls, showing the destination and the number of message units charged for each call.

9. *Telco's business office records.* In the few localities where telco does not give more toll information than the date, city called, telephone number called, and cost, the administrator can secure further information by asking the business office to furnish copies of the stub end of the long distance toll statement. The stub gives the length of the call, the time the call was placed, its type and class, and the telco operator ticket number from which the call is billed. Thus, by looking up the ticket, telco can give further information on third-number, credit card, collect, and person-to-person calls. Telco's record is used by telco personnel when you ask for identification of the person who places a person-to-person collect call to you, a station collect call, or an outgoing person or operator-handled call.

10. *Telco's punched cards and magnetic tapes.* Full information on all toll call details is contained on telco's cards and tapes. If you are experiencing difficulty in securing computer-originated listings of toll calls from telco, you can arrange to rent or purchase the cards and tapes. Your own data processing programmers can then arrange the data in whatever sequence or classification you require for your study.

11. *Equipment inventory forms.* It is essential that the administrator know the details behind his billing for "local service." The equipment inventory form is available from telco's business office on your request. It shows a listing of all charges for your PABX system, including power, switchboards, and switches. The cost of all peripheral key system equipment may be listed under each station number, or it may be added under each key system number at the end of the inventory. All trunks and tie-line charges are itemized, plus all miscellaneous equipment items.

12. *Key system diagrams.* Almost every telco keeps an in-house record of each key system so that the telephone installer can trace all related telephones within a specific key system. This information is very helpful if you decide to charge each department for the exact cost of providing telephone service. Telco will usually make copies of the key system diagrams available to you for study. As changes are made by telco's installer, the key system diagram is updated. You may wish to follow a similar updating process as part of your unit cost-accounting effort.

13. *TWX toll statement.* Western Union sends a list of all TWX connections each month. The date, TWX number called, number of minutes, and cost are shown.

14. *TCS statement.* Western Union sends a detailed billing for the use of its TCS (see Chapter 14). Each individual message sent

via computer over Telex, TWX, Mailgram, or PMS is shown and priced, thus giving valuable information which can be used for unit accounting if desired. The computer access format likewise permits you to insert a departmental code as a part of the "header," which allows you to sort message copies by department number. Cost can then be inserted after the TCS statement is received.

15. *Vendors' invoices.* If you are using interconnected equipment on a rental basis, you can procure copies of needed invoices for that rental from your accounting or purchasing departments.

## RECORD-KEEPING FOR TELECOMMUNICATIONS EXPENSES

In the next chapter, we will be discussing the organization and management of the telecommunications department. Record-keeping helps you in your managing activity and also in cost accounting. The following list of records is applicable to both this chapter and the following chapter.

### Cost per Call or Cost per Message

If you undertake the cost-cutting program which this book suggests, you will wish to keep the kind of record that will help you appraise the result of your cost-cutting effort. True, your total dollar costs for these services may decrease, but this decrease might come about because of changes in your business volume. Thus, try to calculate your unit cost per call or message. You may wish to express a telephone call cost in terms of cents per minute. If you apply some of the techniques of concise talking mentioned in Chapter 1, your cost per call may show a significant decrease. Or the decrease may come about because you have been successful in teaching your callers to dial directly or to use WATS. Management may be more impressed by your percentage reduction in cost per call or message than by a record of less expense in dollars.

### WATS Talking Minutes

As described in detail in Chapter 3, you must determine the total number of talking minutes per month on each WATS circuit. This record tells you if you have additional WATS capacity on an existing WATS trunk or whether you should change the "mix" of your WATS circuits.

### Circuit Outage Time

Your operators should be asked to carefully record the date, time, and duration of any outage on a dedicated circuit, such as WATS,

employees. This chapter is devoted to the principal factors that contribute to successful telecommunications management.

## SELECTING PERSONNEL

If your organization is spending a substantial amount of money with the communications utilities, you may wish to check the possibilities of employing a telecommunications supervisor who has had specific telecommunications experience or training. If the total yearly cash outlay to the telecommunications utilities for the entire corporation is upwards of $750,000, a person who can bring you a 10 to 15 percent cost reduction on this amount may prove to be a valuable addition to your staff.

If you are spending less than that amount, it is possible that management of the telecommunications function can be one of the duties of a person who has the responsibility for supervising administrative services. Or you may find a competent person in your data processing or financial department. Here are some specific suggestions on recruiting sources.

### Members of Your Own Staff

Almost every organization recruits recent high school and college graduates to ensure that younger staff members are being integrated into the organization. You may find someone in this group who is eager to demonstrate ability to management. If the person is intelligent, tactful, able to organize facts, and likes the challenge of selling new ideas to your staff members, this trainee has the potential for being a coordinator of telecommunications services.

### Employees of the Utilities

Men and women who have been employed by the communications utilities are a good source of possible additions to your professional staff. They have been thoroughly trained in the philosophy of the utilities. If they have flexible minds and recognize that they must reevaluate some of their marketing training, they may prove to be excellent telecommunications supervisors.

### Telecommunications Professionals in Industry

As telecommunications experts are very much in demand, there is considerable mobility among able people who have directed telecommunications activity in major organizations. Professional associations, such as the International Communications Association (ICA) and the Communications Managers' Association (CMA) of New York City, have established lists of men and women who wish to consider offers of employment.

### Employment Agencies

To the best of my knowledge, in 1973 only one agency, Personnel Resources International, Inc., of New York City, specialized in furnishing trained telecommunications people to users of telecommunications personnel. A good employment agency should also be able to help you locate people with telecommunications experience.

## EDUCATION AND TRAINING

The field of telecommunications is changing rapidly, and the rate of change will accelerate. The administrator must keep up-to-date. Here are suggestions on sources of education and training.

### Academic Institutions

Colleges and universities are just beginning to recognize the need for giving professional training in administrative and data telecommunications. For instance, the University of Colorado now bestows a Master of Science in Telecommunications degree. Check with your neighboring institutions to determine what is offered. Engineering schools may have helpful courses.

### Publications

In the annotated bibliography you will find suggestions for a basic telecommunications bookshelf, as well as a list of magazines and newsletters. Your librarian can give you the names of many books in the field of data communications. You should also ask the major publishers of business books to add your name to their mailing lists for announcements of new publications in technical fields.

### Meetings and Seminars

Many professional organizations sponsor telecommunications seminars at their annual meetings or at special meetings held in various cities through the year. The Administrative Management Society (AMS) and the American Hospital Association have held such meetings. Your public library can furnish a list of professional associations.

Three organizations regularly schedule meetings devoted to the study of administrative and data communications. You can secure details by writing to them:

American Management Association
135 West 50th Street, New York, N.Y. 10020

Executive Enterprises, Inc.
10 Columbus Circle, New York, N.Y. 10019

Institute for Advanced Technology
Control Data Corporation, 5272 River Road, Washington, D.C. 20016

The meetings sponsored by Executive Enterprises are for one or two days; the other two organizations give three- and five-day courses.

The Bell System has organized seminars, free of charge, for its customers. The meetings are held in specially designed facilities located in New York City, Chicago, and Cincinnati; reservations must be made through the local utility's marketing department. Several different kinds of seminars are offered. Executives may attend a one-day session; administrators may attend a five-day course.

The magazines serving the telecommunications field usually publish a list of future meetings and seminars. *Telephony* and *Business Communications Review* feature quite complete listings.

### Professional Organizations

The manufacturers of telecommunications equipment exhibit their wares each year at the annual meetings of the ICA and the United States Independent Telephone Association (USITA). The members of ICA are the telecommunications administrators of our nation's largest users of these services. Membership is by invitation only; a present member must recommend the election of a prospective member. While nonmembers cannot attend the seminars and business sessions sponsored by ICA, it may be possible for the administrator to secure an invitation to visit the exhibits held in conjunction with the annual meeting. You should try to get acquainted with a member of ICA in your vicinity; he can advise you on the possibility of attending the exhibits.

Similarly, USITA may permit a nonmember to inspect the manufacturers' displays. A registration fee may have to be paid. Details will be found in *Telephony* and *Telephone Engineer and Management.* Manufacturers of interconnected equipment who are exhibiting may likewise extend invitations to prospective customers.

### Local Associations

In major cities, business users of telecommunications facilities may form a local association which meets monthly or bimonthly to exchange know-how and information. Such organizations exist in Chicago, New York City, Ft. Wayne, Cleveland, Toledo, and many other cities. You may be able to secure information by becoming acquainted with the telecommunications administrator of a large corporation in your area.

## UTILIZING SPECIAL SERVICES FURNISHED BY THE UTILITIES

The major telephone utilities offer valuable special services to their customers. These services help you appraise the caliber of your telephone service to your callers. They assist you in determining your switchboard labor requirements. The services may supplement your organization's marketing effort. Here are some descriptions of the services which are quite uniformly available.

### Switchboard Service Observing

The telephone industry has developed time standards for operations performed by switchboard operators. Through application of these standards, operator labor requirements can be determined for each hour of the day. See the section "Determining the Number of Switchboard Positions" in Chapter 10.

Also, the observing telco employee checks the dexterity of the operators. Is the operator using one hand or both hands? Is she using a headset, or is she using one hand to hold a telephone receiver to her ear? Are calls plugged in without an immediate response being given? A plug-in occurs when an operator places a cord in a jack, terminating the ringing signal. The caller may think he is disconnected unless the operator responds immediately. A careless operator may plug in merely to stop the buzzing of her switchboard. This practice may lose calls. The telco observer will teach the operator to answer the call immediately. Then, if other trunks are ringing, the first caller may be asked to wait a moment while the operator reassures other incoming callers. The calls are then connected in the same sequence that they were answered. A trained telco observer knows good switchboard operating practice and can often speed switchboard service materially. As this service is complimentary, it should be used routinely once or twice a year to ensure that you are giving a proper grade of switchboard service to your callers.

### Call-handling Observation

In some Bell System companies, another free service is offered to help you appraise the kind of service your switchboard operators and station users are giving to those who call you. Telco operates an "observing switchboard" which is located on their premises. At intervals, perhaps every six months, the special switchboard is connected to your incoming trunks. A trained observer listens to a sample of perhaps 50 calls. Her listening begins when the main switchboard number begins to ring. She records the number of rings before your operator answers. She listens to the type of answer response given by your operator. She then times the number of rings needed before your station user answers his call once your operator has connected the call to his station line. Finally, the observing operator notes the type of response made by the station user. Did he say "hello"? Or did he answer with his name and/or department? Other observations are made in addition to those itemized above.

Telco's service representative summarizes the results and compares them with the norms established for good telephone service. A written report is prepared and is usually delivered to management with recommendations for action or additional training. We think this free service is an excellent public relations tool for any user of telephone service.

### Traffic Studies

Telco's traffic studies help you make sure that your equipment is adequate for the quantities of calls and conversation minutes which it must accommodate. For information on the subject see the section "Securing Detailed Traffic Studies" in Chapter 10.

## OBSERVING USAGE OF FACILITIES

The competent telecommunications professional does not rely solely on written facts and figures. He *looks* at the operations in his office and in the offices of other companies. Here are a few suggestions of the types of observation you can make.

### Study Requested Equipment Changes

If a department head requests a change of service, the administrator recognizes that he has an opportunity to view how the requested change may affect the operations of that department. It is possible that limited technical knowledge causes the supervisor to order equipment which is needlessly expensive in comparison with other equipment which will provide the same function. Any change ordered gives the administrator a splendid opportunity to inspect *all* the departmental facilities; then he can suggest ways to reduce cost and improve service. Most budget-conscious department heads welcome a visit from a staff expert who can reduce costs.

### Look and Listen at the Switchboard

A second area of observation is that of switchboard operation by one of the executives of your organization. We suggest that the executive spend 30 minutes on a busy Monday morning to observe switchboard traffic at approximately 11:00 A.M. He should wear an observing headset, which can be plugged into a switchboard position. He hears the request of the caller and the response of the operator. The executive listens to typical interrogations which the operator must conduct to route a call correctly to its destination. He may hear requests made for personal service which have nothing to do with company business. He watches while calls go unanswered because some executive insists that the operator perform a function which could have been performed by the executive's personal secretary. Result: the observing executive has a new appreciation for the importance of the switchboard as a focal point of his organization. He also begins to recognize that telephone operators and telecommunications supervisors deserve compensation in keeping with their very important responsibilities.

The administrator should also regularly observe operation of his switchboard at busy hours. He should count the seconds until incoming trunk lamps are extinguished, indicating that an operator has answered an incoming call. He should ask himself if the message-taking function might be better concentrated at an off-switchboard position connected to the switchboard by a special "hot line." You will find that your senior switchboard operator will be able to give you a great deal of information about the problems encountered by your customers in trying to reach you. She will also tell you about the "long talkers" who are monopolizing your expensive dedicated line service. And you'll learn a lot about the volume of incoming personal calls destined for your employees.

### Observe Teleprinter Operation

A third point for observation is the message center where your teleprinters are located (see Chapter 14). You should select typical outgoing and incoming messages. Inspect them for brevity, clarity, and mode of transmission. Are your people using abbreviations to conserve key strokes? On an incoming transmission, is the teleprinter machine printing very slowly and hesitating? This indicates that some distant operator does not know that her message should have been prepared off-line on tape and then transmitted at the full Telex speed of 66.67 words per minute or the full TWX speed of 100 words per minute.

### Inspect Your Equipment Room

A fourth area of observation is the room which houses the equipment which operates your PABX system. Is the room being kept clean? Dust can really interfere with telephone switches. Have the meters attached to the racks been read regularly? (See Chapter 10.) Your visit to the equipment room may also help you get better acquainted with your telephone repairman. He can teach you how to make a switch "busy" so that you can observe how the loss of that facility affects your meter readings. Such an experiment may help you decide whether disconnection of switching equipment will reduce cost.

### Visit Other Offices

You can also do some observing of the way other telecommunications people run their departments. As you know, there is a scarcity of how-to-do-it literature in this field. Thus, the people trying to manage the telecommunications function have had to do some original thinking on methods and means. You'll find that your counterpart in other organizations will be delighted to talk shop. You'll be given a complete inspection of facilities. Your friend will be proud of the control techniques which have been developed to get the job done. You'll find that such exchange visits to other organizations will provide you with fresh ideas. Then, as your acquaintanceship spreads, perhaps many of you will form a local association. Such groups can be very effective in dealing with your local utility. If a new tariff is proposed which may work to your disadvantage, group action can be taken with the utility and with the state commission which governs the utility.

## RECORD-KEEPING

In Chapter 16 I listed certain types of records which will help you in your effort to account for telecommunications costs. Here are suggestions on records you need for better management of the telecommunications function.

### Numerical List of Assigned Station Numbers

While your intraoffice telephone directory will locate your people by telephone number, you also will need a list which will tell you to what person or department a specific line has been assigned. Be sure to keep your numerical list up-to-date as you change telephone number assignments.

### Numerical List of Restricted Telephones

As job responsibilities change, and as a telephone number is assigned to a different person from the one originally in that office, you may receive requests to change a station line from one class of restriction to another, or you may be asked to unrestrict a line. Perhaps a person having an unrestricted line is reclassified into the restricted category. Your numerical list of all restricted lines, showing the class of restriction, will help you regularly evaluate whether a particular line should continue its restricted classification or whether the line user's constant requests to the operator for an outside line may justify consideration of a service change.

### Record of Meter Readings

In Chapter 10 you learned how meter reading can help you evaluate the service given by your PABX system. A loose-leaf book which shows your weekly or monthly meter readings will give you a valuable history of your system's traffic patterns.

### Visible Index of Personnel

Your switchboard operators need a fast, easily read index of all personnel for whom calls may be received. A loose-leaf notebook is not satisfactory, as it cannot be kept up-to-date easily, and constant reference will soil and tear its pages. All manufacturers of visible index devices sell indexes designed especially for installation between switchboard panels. Other indexes may be mounted on pedestals above the switchboard or beside the console. The indexes usually feature movable cards mounted vertically so that new names may be placed in the correct alphabetical order. Sometimes index cards are mounted on wheels which can be turned quickly to the desired spot.

### Record of Hiring, Terminations, and Transfers

There *must* be close liaison between the personnel or payroll department and the switchboard supervisor. It is embarrassing for the operator to report to a caller that there is no record of a person when the name is not shown on the visible index. Such calls must be relayed to the personnel department, delaying the business purpose of the call. This problem can easily be solved by convincing the personnel director or payroll supervisor to give a copy of the written notice of a hiring, transfer, or termination to the switchboard supervisor.

### Listing of Employees without Telephones

When an organization employs a large clerical staff, many calls will be received for employees who do not have a numbered station line assigned to them. If the organization's policy permits connecting such calls to the department head's telephone or to a telephone near the employee, the call connection can be expedited by the alphabetical listing showing the employee's department location.

### The Triple-Number Directory

Most intracompany telephone directories show an individual's or department's name and a single telephone number. If the telephone system is designed so that a call to a person automatically rotates to another employee when the first line is busy, the caller at least has a chance to leave his name and number with the next person in the hunting sequence. However, after a period of time, many hunting sequences are changed or canceled entirely. If a line is busy, the calling person must call again.

To help solve this problem, a number of organizations use a double- or triple-number system. The listing might look like this:

328—Morgenthaler, David
362— Secretary, Lyn Cameron
219— Messages: 9th Floor Reception

Or, to conserve space, instructions can be given at the beginning of the directory to call the numbers given in order. The caller assumes that the first number is the person he wants; the second number is a person who is related to the job function of the executive; the third number is a continuously manned clerical or secretarial position where a message may be taken and delivered. Such a listing might read:

328-362-219 Morgenthaler, David

Please note that the number precedes the name, but the list is in alphabetical order, not numerical order. This is because a new switchboard operator can learn numbers more quickly in this format, and the operator can also procure a number faster from the visible index when the number is linked directly to the surname rather than being placed at the end of the copy line.

## STANDARD SETTING

If you wish to change the atmosphere in your organization from "permissive" to "controlled" (see Chapter 16), you should consider setting standards which may not be exceeded by telecommunications users without executive approval. Here are some examples of the standards you might set.

### Long Distance Toll Calls

When a toll call must be made, it must be directly dialed without the assistance of the utility's operator. Credit cards must not be used at any point where direct dialing facilities exist, such as an employee's residence or one of your organization's branches.

### Message Units

Clerical personnel will not be allowed to make local calls unless the department head certifies that this is part of a specific clerk's job responsibilities.

### WATS Lines and Other Dedicated Circuits

All those wishing access to these lines will make their request by dialing the reservation operator. No priorities will be given: all calls will be handled in the order in which the reservation is received. If a call is urgent and cannot be delayed for connection to a dedicated circuit, it will be awarded a toll circuit by the operator, and the charge for that call will be billed against the cost center of the originator.

### Rental of Telephone Equipment

Any request for a change, move, addition, or removal of equipment must be approved by the telecommunications coordinator. His decision on the type and quantity of equipment ordered will be final. In general, no line illumination will be permitted on key systems of two lines or less. Hands-free telephones, when authorized, will be furnished by outside vendors. Dial intercoms may be used only for call distribution by a departmental secretary.

### Transmission of Domestic and International Wire Messages

The originator of the message will check the appropriate box on the message blank indicating whether the message is to be delivered during the present business day or on the following business day. The message center supervisor will have the authority to choose the carrier and route for handling the message and to reject a message which is unclear or does not meet the message center's concise writing standards. The supervisor will gladly instruct the originator on message writing techniques applicable to domestic and international messages, as explained in Chapter 14.

## EVALUATING YOUR ACCOMPLISHMENTS

When you have instituted the controls and methods which you believe will improve service, reduce costs, and enhance public relations, you will want to make some kind of evaluation of the results you have achieved. In Chapter 16 I suggested that you keep records on cost per call and cost per message. These unit costs should definitely decline as you implement the suggestions given in this book. You might also consider establishing a few additional indexes.

### Percentage of Telecommunications Expense to Sales

This figure has been used in some of the magazine articles written about telecommunications. Organizations that use the telephone extensively in sales solicitation may have much higher costs than companies that do not do this. If your costs exceed 1 percent of your company's net sales figure, you should find out *why,* as the percentage in most organizations is much less.

### Cost per Telephone Instrument

This figure is useful in comparing the costs of one department with another department when you operate in a permissive atmosphere. You divide the total number of telephone lines into your total telephone bill for each department. If you work hard on reducing the expense of key systems and peripheral equipment, your cost per telephone will show a significant decrease. Further, if you make this analysis on a departmental basis, and if there is rivalry among your department heads on budget performance, the publication and distribution of this statistic may help you secure better cooperation on cost-reduction efforts.

### Cost per Minute

This statistic, applied to toll, WATS, and other dedicated services, will help you evaluate your success in convincing people that they should use the wholesale services.

### Examples of Typical Records

I am grateful to Joseph Lawless, Manager—Teleservices for Nabisco, New York City, for permitting me to reproduce some suggested forms for recording the information we have discussed. The forms are shown here as Tables 1, 2, and 3 and Figures 1 and 2. The dollar amounts given in the examples are not Nabisco figures; they are used only for illustrative purposes.

The form shown as Table 1 summarizes expenses for several

**TABLE 1 Domestic Annual Communications Expense**

| Department name | Year | Cost of equipment rental | Cost of local and long distance calls | Total cost of voice communications | Cost of written communications | Total cost |
|---|---|---|---|---|---|---|
| General office | 1970 | $125,073 | $134,154 | $259,227 | $142,886 | $402,113 |
| | 1971 | 150,421 | 175,186 | 325,607 | 156,229 | 481,836 |
| | 1972 | 175,398 | 198,827 | 374,225 | 165,783 | 540,008 |

years by department to help top management secure cooperation from department heads on meeting budget performance. The form continues with a listing of other departments and expenses. The company-wide totals are given at the end of the form.

**TABLE 2 Monthly Cost of General Office Telephone Service**

| Department name | % instruments in department | Number instruments in department | Instrument rental | Departmental % of total cost | Average cost per instrument | All other telephone equipment | Departmental % of total cost | Total cost |
|---|---|---|---|---|---|---|---|---|
| Accounting | 1.4 | 12 | $35.65 | 1.0 | $2.97 | $45.48 | 1.2 | $81.13 |

The form shown as Table 2 helps stimulate competition between departments to keep costs low. The complete form lists all departments. At the bottom of the form are the average cost per instrument, the average cost per department of all other telephone equipment rental, and the monthly average for each department of total cost. The figure for the cost of all other telephone equipment is obtained by taking the percentage of total instruments contained in the department and multiplying it by the *total* cost of all other telephone equipment for the office as a whole.

Another form, shown as Table 3, helps the administrator study yearly changes in cost levels.

As in any form of statistical presentation, a graph can get the message across more dramatically than a column of figures. Figure 1 helps the administrator see how his yearly costs climb in relation to his company's sales figures. Figure 2 (on p. 294) shows a tele-

**TABLE 3 Allocated Cost per Telephone—General Office**

| | Sept. 1969 | June 1970 | May 1971 | May 1973 |
|---|---|---|---|---|
| Cost of local lines, both in and out, telephones, and switching equipment | $7.07 | $7.78 | $8.39 | $10.11 |
| Cost of message units | 2.37 | 2.24 | 2.52 | 3.48 |
| Cost of installation, moves, and changes | 0.99 | 0.98 | 0.36 | 0.81 |
| Cost of New Jersey FX calls | 1.35 | 1.57 | 1.51 | 1.69 |
| Cost of long distance calls charged to the main number | 0.74 | 0.22 | 0.63 | 0.79 |
| Local and federal taxes | 0.74 | 0.46 | 1.10 | 1.03 |
| Total cost of instrument | $13.26 | $13.25 | $14.51 | $17.91 |

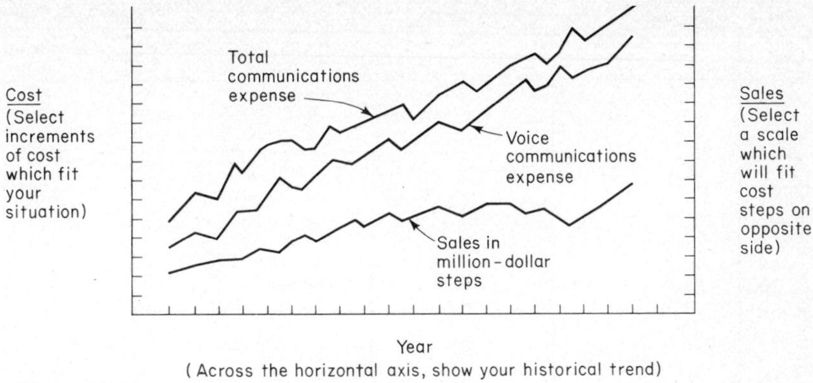


**Figure 1.** Annual costs versus sales.

phone user how he can save money for his company by agreeing to a delay in placing his long distance call so that it can go through one of the company's dedicated circuits rather than be directly dialed on the long distance network.

## EVALUATING SPECIAL SERVICES

Outside organizations will be sending you literature on special services they can provide to assist you in your management and training responsibilities.

### The National Account Concept

AT&T provides special assistance to large organizations which use dedicated private line voice networks or similar expensive services which connect many branches with headquarters. If you are designated by AT&T's Long Lines department as a "National Account," a specific individual will be assigned to your organization on a full-time or part-time basis. The person may be a member of the Long Lines staff or a staff member of the utility which serves you. That individual takes the responsibility for developing all the information you need to help you solve a problem arising at a distant location. Equipment listings or billing copies can be procured. The National Account representative keeps in contact with the marketing representative serving your office in the distant city. Coordination is provided, and your job becomes much easier. Your local marketing representative will be glad to contact the nearest AT&T Long Lines office to determine whether your organization is eligible for National Account status.

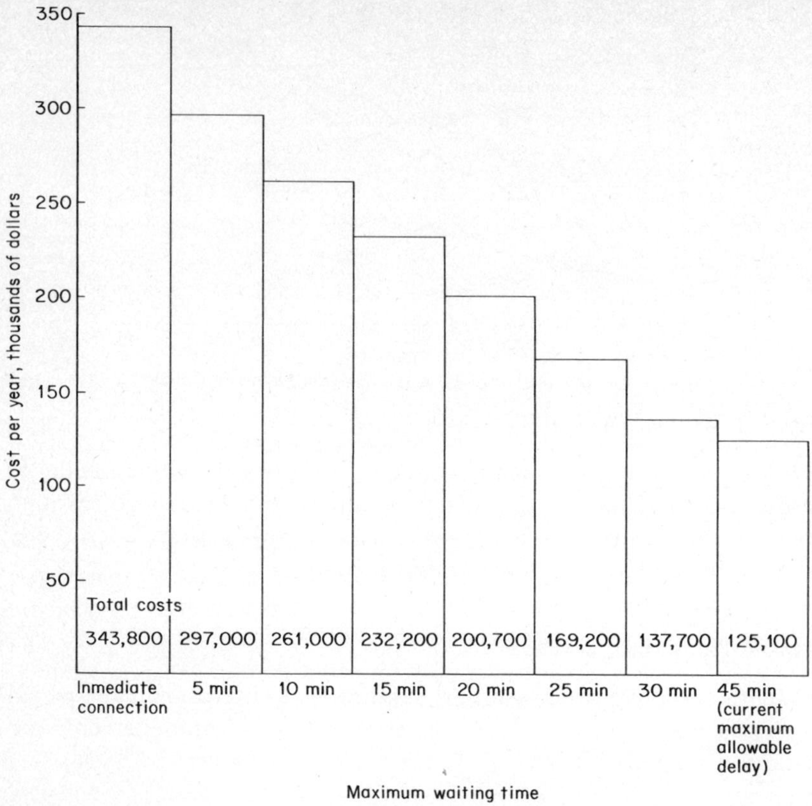


**Figure 2.** Effect of delay on cost of dedicated circuits.

## The "Phone Power" Program

AT&T has developed an excellent program which trains your staff members on how to do business by telephone. Most major cities have a "Phone Power" specialist. That individual comes to your office for conferences with you and your marketing staff. Much written material is made available to you. There is no charge for this service.

## Training Programs for Employees

Many outside organizations advertise do-it-yourself training kits which use workbooks and cassette recordings for training in good telephone manners and sales techniques. While it is difficult to tailor a standard program to an organization's individual needs, you may wish to ask the vendor to send you his material for listening and evaluation.

### Procure and Evaluate Product Literature

The bibliography which follows this chapter lists magazines that carry information which may be helpful to you. Several of the magazines contain detachable cards on which you may indicate your interest in the products or services advertised. You can secure many excellent ideas by keeping up-to-date on these technological developments.

Also, every Bell System subsidiary company has access to AT&T's excellent product and service publications. In addition, the local utility may publish its own descriptive literature. In most telco marketing offices these bulletins are displayed in a rack. The marketing representatives can pick up copies when they are about to consult with a customer about FX, tie lines, WATS, switchboards, Centrex, or supplementary equipment for special purposes. I suggest that you ask your marketing representative to let you inspect his literature rack. The publications are available to you at no charge. They will help you understand the equipment and services offered by the utilities.

### Evaluate Present Personnel Department Indoctrination Work

Make sure that your personnel department is instructing new employees on the workings of your telephone system. Such training is particularly important if you are using WATS or other dedicated facilities. If you wish to emphasize telephone courtesy, you may want to develop a set of standard secretarial answer responses, such as:

> "Mr. Jones's office, Miss Smith speaking." "I'm so sorry, sir, but he's in conference now. I'll be glad to try to interrupt him if your message is urgent, or I can have him call you when he's free."
>
> "I'm sorry, sir. He's away from his desk at the moment. I expect him back around two o'clock this afternoon. May I have him call you, or is there something I can do to assist you now?"

You can think of many more standard situations for which you could develop courteous, enthusiastic responses which could be taught to all your people.

### Take Advantage of Telco's Training Services

All major telcos offer training programs for both new and old employees. Movies are shown on telephone courtesy. Recording equipment helps operators and secretaries hear their telephone voices. Many written publications on telephone courtesy are made available to you. You should send new switchboard or console operators to telco for training.

## EVALUATING THE USE OF OUTSIDE CONSULTANTS

In recent years there has been a rapid increase in the number of firms and individuals who represent themselves as communications consultants. The yellow pages of your telephone book will probably list a great many firms which would like to assist you in running your department. There are several different approaches to consulting.

### Contingent Consulting

In effect, your prospective consultant says: "Don't pay me a cent unless you actually accept my recommendations. Then pay me a percentage of the savings which you realize." This type of consulting is very common, but it carries many risks.

Such consultants usually propose an item-by-item inventory of your equipment, so the inventory may be compared with the equipment listing supplied by your utility. If you are being billed for equipment which is not in use, a request for an appropriate refund is made on a retroactive basis. You then share the refund with the consulting organization.

However, it is not necessary for you to employ a consultant to do a detailed inventory. Instead, your consultant, working with your utility's marketing representative, should train members of your staff in the meaning of the codes appearing on your equipment inventory record furnished by telco. You learn how to read a key system diagram. Working under the consultant's direction, you may inventory a friendly department to get the "feel" of what must be done. Then, at your leisure, you conduct the inventory yourself. Or, if you wish, telco will make the inventory for you at no charge. One of your staff members, or perhaps a member of your internal auditing staff, can participate. If discrepancies are found, you file your own request for a refund and keep the entire amount. If you find equipment present for which you are not being billed, you notify telco, and they will begin billing you for that equipment.

Contingent consultants usually ask their client to sign an authorization which permits the consultant to deal directly with your utility. If the consultant decides that a particular piece of equipment should be disconnected, that order can be given to the utility without your approval. As the consultant's compensation is directly related to savings produced by the disconnection, the consultant may be overly enthusiastic about eliminating equipment which you need for proper operation of a particular department. And, because installation charges have doubled and tripled in recent years, you may find it very expensive to replace the disconnected equipment. However, most contingent consultants make sure that their customer signs a legal document which ensures continuing revenue to them on the original disconnection, regardless of changes which their client must make later.

As you can readily understand, both telcos and equipment users have experienced difficulties in negotiating with consulting firms which make their compensation contingent on savings. Many lawsuits have been filed to avoid liabilities imposed by contingent contracts. Further, the Association of Consulting Management Engineers (ACME) does not permit its members to engage in contingent consulting, which is considered an unethical trade practice in the management consulting profession. ACME publishes a booklet which contains information on securing the best possible results from management consultants. It is available free from ACME, 347 Madison Avenue, New York, N.Y. 10017.

### Securing Consulting Assistance in Evaluating Your Performance

Consultants who follow the ACME ethical code are available to you on a fee basis which is related only to the professional time which is consumed in your offices and in the offices of the consultant. Living and traveling expenses are billed separately. The consulting firm makes no charge for a preliminary conference with you, either in person or by an exchange of correspondence. Most consulting firms also offer to examine your cost data and equipment listings without charging for the work involved. The consulting firm then sends you an informal proposal, which lists the work to be done, states the approximate time that will be consumed, and describes the amount of time your staff members should devote to the collaborative project. A firm or "upset" price for the engagement is quoted. The proposal usually indicates that careful time records will be kept on each assignment. If the total time expended is less than the time estimated, the client is billed at the smaller amount. Most consulting firms following the ACME code do not require their client to sign a formal contract. Instead, a simple letter of acceptance is sent to the consulting organization, and work is then scheduled.

In considering proposals from consultants, remember that the essence of good management consulting is *teaching*. There is no need for a consultant to perform work which your own staff can perform once you understand what must be done. The consultant diagnoses problems, conceives solutions for them, teaches your staff members how to proceed, and then provides general supervision from time to time. The trained consultant knows what services can be provided by the utilities to their customer without charge and will not undertake to perform services which the utility can perform equally well. Instead, the client, the consultant, and the utility's marketing people all work together to bring about the desired result.

In selecting your consultant, be sure to ask for information about the background and experience of the consultant's staff member or members who will be assigned to your engagement. Try to meet those individuals in advance of the assignment so you can make your own judgment of their competency. You should secure information on the background of your consulting firm's senior principals. Equally important, the consulting organization should furnish references to you. Be sure that you are given the name of a top officer of the consultant's client company so that you can make a direct telephone call to that individual. You should check at least three references, making certain that the work performed for the organization you call is similar to the work you wish performed for you.

If you choose your consultant properly, the resulting professional relationship may be very helpful to you over a period of years. A good management consultant is very much interested in broadening your personal knowledge in your chosen field. Your job becomes more interesting. And because a properly trained management consultant has usually been an executive in a major organization, he can help you in your contact with top management. Often, suggestions made by a consultant are implemented much more quickly than the identical suggestion made by a staff member. Also, it is quite possible that your collaboration with a good consultant can save you weeks and months of personal research and, in addition, increase your own knowledge, prestige, and compensation.

## REPORTING TO MANAGEMENT

Now that I have mentioned prestige and compensation, I will close this book with an urgent plea to the telecommunications administrator and an important message for the executive who takes responsibility for this area of administrative management.

To the telecommunications administrator: *Inform your management of what you are doing!* Keep records of the savings you have made. Draw graphs of telecommunications expense in relation to sales volume. Show graphically the decline in cost per call, cost per message, or equipment cost per telephone. Ask your supervising executive for at least two conferences a year to discuss what you have done and what you plan to do or what management wishes you to do. These efforts should be very effective in convincing management of your professionalism. Your compensation should increase as management recognizes the importance of your work to the organization as a whole.

Also, *study data communication.* Our computer-controlled world is constantly requiring more and more telecommunications facilities for the transmission of data. If your responsibilities are chiefly those involving administrative messages, you may wish to seek additional responsibility for sending your company's data to and from your computer. Many data processing experts are attending seminars on administrative telecommunications so that they will understand *your* responsibilities. It would be wise for you to start to learn about the needs of your data processing people. Books and seminars abound in the data communications field.

To the responsible executive: *The person who manages your telecommunications services holds a very important position.* This is because your telephone and written wire message services are vital components of your competitive existence as an organization. May I suggest that you try an experiment as you travel? Call your main switchboard number and pretend to be a prospective customer. Purposely, tell the answering person that you're not sure about the identity of the individual or department concerned with your interest or problem. Then start observing. Did your switchboard operator answer promptly? Did you have a long wait before the department to which your call was directed answered? Was your call transferred several times? Did those talking to you have good voice inflections? Did they sound as if they were *glad* to be talking to a prospective customer? You will quickly decide if action needs to be taken.

Here is an important question for the responsible executive. Do you know your company-wide total cash outlay for all telecommunications services? You may be surprised to find how much you are spending with the communications utilities. Have you given one of your staff members authority to supervise and control these expenses at both headquarters and your branch locations?

Many cost-conscious companies permit regional executives to make minor changes in their use of telecommunications services. However, if an entire system must be changed or a new service or system installed, the regional manager is either encouraged or required to consult the individual who is the headquarters staff expert in this fast-changing technological field.

In turn, the executive supervising that person must recognize that the telecommunications coordinator should be given direction. The coordinator should be encouraged to attend seminars and conferences which will help him keep up-to-date. The executive should set goals—and should confer regularly with the coordinator to measure progress. If you believe that your organization can benefit from more thoughtful and enthusiastic handling of incoming calls and messages, give your coordinator those objectives. There are many free or inexpensive training services which can be utilized. A good telecommunications manager will perceive that the financial health of his company depends on the enterprise's *customers,* and he will do all he can to see that they get prompt, careful, and courteous attention.

The man or woman who supervises your telecommunications services is a valuable member of your management team. This individual can enhance your organization's public relations image and, even more important, bring about a most impressive cost reduction if given the authority to set standards and supervise the use of your facilities. The telecommunications expert deserves compensation in keeping with the professional ability required by this responsible position. Select the right person, pay him or her well, and you'll be greatly rewarded for having taken an interest in this fascinating field!

# BIBLIOGRAPHY

## Books

Brackett, Gilbert R.: *Telephony's PBX Service Adviser's Handbook,* Telephony Publishing Corp., 53 W. Jackson Blvd., Chicago, Ill. 60604, 1967, 132 pp. Designed to train the service representatives of the telephone utilities, this publication is helpful to the telecommunications manager because it explains how good service can be given to those calling your main switchboard number.

———: *Telephony's Telephone Traffic Engineering Handbook,* rev. 2nd ed., Telephony Publishing Corp., 53 W. Jackson Blvd., Chicago Ill. 60604, 1967, 226 pp. This book provides brief explanations of traffic engineering principles and their application to step-by-step switching systems. Mathematical tables used by telco's traffic engineers are included.

Gentle, Edgar C., ed.: *Data Communications in Business,* American Telephone and Telegraph Co., New York, 1965. The book must be ordered through Publishers' Service Co., Room 600, 75 Varick St., New York, N.Y. 10013. This publication is directed to those who wish to begin a study of data communications. It contains helpful illustrations and is written in easily understood language.

Greenfield, Harvey, and Frank K. Griesinger: *Sale-Leasebacks and Leasing in Real Estate and Equipment Transactions,* McGraw-Hill Book Co., 1959, 106 pp. This book is out of print. A paperback edition produced by University Microfilms is available from Frank K. Griesinger and Associates, Inc., 838 National City Bank Bldg., Cleveland, Ohio 44114. This book is helpful to the reader who is studying whether to purchase or lease interconnected equipment.

Smith, Emerson C.: *Glossary of Communications,* Telephony Publishing Corp., 53 W. Jackson Blvd., Chicago, Ill. 60604, 1971, 547 pp. The technical terms used in the telecommunications industry are defined in this publication.

Stone, Roy: *Cost Reductions in Wire Communications,* rev. ed., Christopher Publishing House, Boston, 1959, 350 pp. This book is out of print. A paperback edition produced by University Microfilms is available from Frank K. Griesinger and Associates, Inc., 838 National City Bank Building, Cleveland, Ohio 44114. Colonel Stone is a retired Bell System engineer. His book was the first text written on administrative communications. It contains splendid material on determining trunk and switch requirements for telephone systems, plus careful explanations of meter reading, switchboard work coefficients, and similar subjects.

## Magazines and Newsletters

*Business Communications Review* (bimonthly), P.O. Box 32, Downers Grove, Ill. 60515. A well-written, scholarly review of developments in telecommunications, it includes a listing of current seminars being held on this subject.

*Communications News* (monthly), Brookhill Publishing Co., 402 W. Liberty Drive, Wheaton, Ill. 60187. This tabloid-size magazine features new products and case histories of problems solved by telecommunications administrators.

*Exchange Service Telephone Rates* (published irregularly), National Association of Regulatory Utility Commissioners, P.O. Box 684, Washington, D.C. 20044. This booklet lists flat- and measured-rate trunk charges for major cities in each state.

*Inter-connection* (monthly), Interconnection Publications, Inc., P.O. Box 1535, S. Hackensack, N.J. 07606. Current developments in the interconnection industry are covered in this lively newsletter.

*Long Distance Message Telephone Rates* (published in the fall of each year), National Association of Regulatory Utility Commissioners, P.O. Box 684, Washington, D.C. 20044. This helpful booklet lists all intrastate long distance rate schedules. In addition, tables are included for long distance interstate calling and for calling between the United States, Canada, and Mexico.

*Telecommunications* (monthly), 610 Washington St., Dedham, Mass. 02026. This magazine deals with the technical aspects of telecommunications.

*Telecommunications Reports* (weekly), 1204 National Press Bldg., Washington, D.C. All rate and tariff matters brought before the FCC and the state commissions are thoroughly covered in this publication.

*Telecommunications Week* (weekly), 1204 National Press Bldg., Washington, D.C. This is a one-page condensation of *Telecommunications Reports*.

*Telephone Engineer and Management* (twice monthly), Brookhill Publishing Co., 402 W. Liberty Drive, Wheaton, Ill. 60187. A magazine serving the independent telephone utilities, it features articles on current technical developments.

*Telephony* (weekly), Telephony Publishing Corp., 53 W. Jackson Blvd. Chicago, Ill. 60604. The leading magazine covering the independent telephone industry, it features technical articles, lists meetings and seminars, and carries columns reporting both federal and state regulatory activities.

# INDEX